Competitiveness in
International Food Markets

Competitiveness in International Food Markets

EDITED BY

Maury E. Bredahl, Philip C. Abbott, and Michael R. Reed

Routledge
Taylor & Francis Group
LONDON AND NEW YORK

First published 1994 by Westview Press

Published 2018 by Routledge
52 Vanderbilt Avenue, New York, NY 10017
2 Park Square, Milton Park, Abingdon, Oxon OX14 4RN

Routledge is an imprint of the Taylor & Francis Group, an informa business

Library of Congress Cataloging-in-Publication Data
Competitiveness in international food markets I edited by Maury E.
  Bredahl, Philip C. Abbott, and Michael R. Reed.
    p. cm.
  Includes bibliograph~cal references and index.
  ISBN 0-8133-1736-3 /
  1. Produce trade. 2. Produce trad—overnment policy.
I. Bredahl, Maury E. II. Abbott, Philip C. (Philip Chase), 1949–
III. Reed, Michael R.
I. Danopoulos, Constantine P. (Constantine Panos)
HD9000.5.C6265 1994
382'.415—dc20   93–42931
                                                                      CIP

ISBN 13: 978-0-367-00773-7 (hbk)

In Memory of

Jason E. Bredahl

*How can the whole world seem so empty when
only one person is missing?*

Contents

PART THREE
CONCEPTUAL FOUNDATIONS AND ASSESSMENTS FROM TRADE AND MACROECONOMIC THEORY

PART FOUR
ASSESSMENTS OF THE COMPETITIVENESS OF NATIONAL FOOD SECTORS

Preface

The International Agricultural Trade Research Consortium (IATRC) is a group of economists from around the world who are interested in fostering research and providing a forum for the exchange of ideas relating to international trade of agricultural products and commodities. Each summer the IATRC sponsors a symposium on a topic relating to trade and trade policy from which proceedings are published. For a list of past symposia and related publications, contact Laura Bipes, IATRC Administrative Director, Department of Agricultural and Applied Economics, University of Minnesota, St. Paul, MN 55108, United States of America.

The editors acknowledge the help of Laura Bipes of the University of Minnesota for making arrangements for the symposium. The financial support of the Economic Research Service, the Foreign Agricultural Service and the Cooperative State Research Service of the United States Department of Agriculture, and Agricultural Canada made the conference and this book possible. A special thanks is extended to Dr. Stanley Wilson, former executive director of the Council on Agricultural Science and Technology, for his support and guidance. Nancy Ottum provided editorial assistance and prepared the final camera-ready copy for this book.

Maury E. Bredahl
Philip C. Abbott
Michael R. Reed

PART ONE

Introduction

1

Introduction

Maury E. Bredahl, Philip C. Abbott, and Michael R. Reed

The papers in this book were originally presented at a symposium sponsored by the International Agricultural Trade Research Consortium: Competitiveness in International Food Markets. The conference, and the selection of papers presented, grew out of our investigation of agricultural and food product high-value-added trade sponsored by the Council on Agricultural Science and Technology (CAST).[1] The project was selected by CAST because of its belief that the United States was performing rather poorly in those markets. Our assignment was to identify and analyze the factors affecting performance of the United States in international markets for high-value-added products.

We received enthusiastic support for the CAST project and the IATRC conference from Canadian colleagues. With the implementation of the Canadian-U.S. free trade agreement and the negotiation of the North American Free Trade Agreement, *competitiveness*, or more correctly, a feared *lack of competitiveness*, has become a Canadian national obsession. That concern is evidenced by the formation of the Agri-food Competitiveness Council, the creation of a Competitiveness Branch in Agriculture Canada, and by a host of government and private sector sponsored studies on competitiveness.

Despite the widespread interest in *competitiveness* in North America, we found, with a few notable exceptions, a curious lack of conceptual models and of empirical studies to synthesize for the CAST project. We found many studies that bemoaned the fact that the unit value of food exports from the United States was about half that of world trade in food products and about one-fourth that of the exports from the European Community. The poor trade performance was attributed to a lack of focus and effort by U.S. agribusiness firms, subsidization by the European Community, giving European firms an unfair advantage, and by farmers producing to the specifications of government programs rather than for markets. Most of these studies favored government programs to expand U.S. high-valued food exports, often through a proposed subsidization of further processing.

Other studies simply dismissed the issue of competitiveness outright. North American farmers were argued to be the most efficient in the world, and given the richness of the natural resource endowment, the exportation of unprocessed bulk commodities was deemed to be natural and to be expected as that is the outcome predicted by comparative advantage. These studies focused on crop production almost to the exclusion of other types of food production, and the judgments were often based on absolute advantage, not on relative costs as dictated by the theory of comparative advantage.

Not only was the received theory misused, and so a number of wrong answers reached, but also we found the underlying theory and models deficient because a number of "real world" trade flows could not explained. Although trade varies from year to year, the United States is the largest exporter of beef *and* the world's second largest importer; the European Community is the second or third largest importer *and* exporter of beef. Taiwan and Denmark, two small nations with few natural resources, are the largest suppliers of pork to Japan, holding over 80 percent of that market between them. The United States and France, two nations with large cereal and oilseed surpluses, have not been important in international pork and poultry markets. Yet, traditional trade theory suggests that nations export those products using their most abundant and cheapest resource.

The second chapter summarizes these issues. It reviews several definitions of competitiveness, the wrong answers found in economic literature and the popular press, and the potential contributions and shortcomings of trade theory. A number of "right questions" are asked, and a continuum of factors affecting competitiveness are identified. A conceptual framework based on economic activities leading to certain end-use characteristics in products, termed "the four economies of agriculture," is developed for evaluating the importance of several factors to agricultural competitiveness.

A basic question in the selection of an operational definition of competitiveness and in the selection of analytical techniques and models to address U.S. or Canadian performance in high-value-added agricultural trade is the appropriate unit of observations. Critics of trade theory point out that firms trade, nations don't; that firms make investment and marketing decisions, nations don't; and that firms compete in international markets, nations don't.

We did find that the business strategy literature developed interesting concepts, however *ad hoc*, that have been largely ignored by all but a small part of the agricultural economics profession. Market structure and firm strategy clearly affect export performance, and nonprice factors may be of greater importance than price in explaining superior performance by some countries in international markets. The contribution of the business strategy literature and the importance of nonprice factors is developed in the second section: *Conceptual Foundations and Assessments from Business Strategies.*

The lead-off paper by van Duren, Martin and Westgren makes an important step forward in integrating the business strategy literature with economic theory and develops a conceptual framework for evaluating sector competitiveness. With that said, the conceptual framework lacks the rigor and mathematical precision favored by most trade economists. The second paper by Allen and Pierson illustrates the range of nonprice factors that can affect performance in any market. However masterful their written presentation, it could not possibly approach their colorful and lively presentation at the conference. The paper departs even further from a stylized conceptual framework than the van Duren, Martin and Westgren paper. Some trade economists will likely find the paper "fuzzy" and of little value in conceptualizing a mathematical or empirical trade model intending to capture the effects of nonprice factors. The final chapter in this section, by Reed, finds evidence that nonprice factors play an important role in explaining U.S. agricultural and food exports using more traditional econometric techniques.

Critics of the business strategy approach point out the very factors that are important in explaining competitiveness are taken as given in business strategy literature. Lau points out that the three principle sources of economic progress are growth of the stock of capital and labor and technical progress. A business strategy approach takes as given the very factors that most trade economists expect to explain. An explicit intent of the conference was to bring these two very different lines of thought together.

The emergence of trade models under the rubric of the "New Trade Theory" and analyses that key on differentiated products, economics of scale, and market power provided the rationale for the next section: *Conceptual Foundations and Assessments from Trade and Macroeconomic Theory*. The intent was to review developments in these areas and in the process to explore the degree of comparability with the explanations of the business strategy approach. The first paper by Ethier reviews the evolution of trade models and the recent development of theoretical models to address product differentiation, direct foreign investment, and technology diffusion.

In a stimulating paper very much in the tradition of macroeconomic analysis, Lau investigates the importance of economies of scale and the growth of productivity in selected developed countries. Sheldon and McCorriston, in the tradition of theoretical trade models, argue that analysis of economic growth requires a dynamic framework and conclude that sectoral market share is not an important indicator of economic growth and competitiveness. Endogenous technical change is seen as key to the evolution of competitiveness. Policy intervention to augment economic growth in some sectors may be beneficial, but an interventionist response to other nations' actions may not be warranted. The final paper in this section by Kalaitzandonakes, Gehrke and Bredahl argues that competition from imports stimulates productivity growth. Empirical evidence of sharply different rates of growth for the Florida winter vegetable industry is presented.

The fourth section, *Assessments of the Competitiveness of National Food Sectors*, draws together several studies of the competitiveness of the food and agricultural sectors of the United States, Canada, New Zealand, and Denmark. The definition of competitiveness varies greatly among the country analyses, but a common theme is the microlevel idea that competitiveness is the ability to profitably gain and maintain market share. McDonald and Lee touch upon many issues the United States is struggling with — trade barriers on agricultural products, value-added versus bulk exporting, and in-bound versus outbound direct foreign investment. The overall theme, however, focuses on U.S. productivity levels versus Brazil and the European Community. Handy and Henderson show the importance of multinational corporations in world food markets and analyze their behavior relative to location of production, product lines, and earnings.

Three papers address competitiveness in Canada. Miner argues that there is a substantial linkage between high border controls (at the federal and provincial levels) and a lack of competitiveness in many Canadian food subsectors. His analysis indicates that Canadian firms are less competitive than their U.S. counterparts and the gap in productivity between the two nations is increasing. Hazledine develops a new measure of competitiveness — market mass — to compare Canadian with U.S. processors. He then links factors associated with high levels of market mass (or competitiveness) and finds that such subsectors are not particularly associated with superior export performance, productivity levels, degree of protection, or industry concentration. The paper, by Brink and Ash, focuses on how the competitiveness issue has shaped policy choices in Canada. Their conclusion is that governments should concentrate on providing a social and economic environment conducive to increased competition and upgrading the capabilities of labor and other resources.

The final two papers concern countries that are very export-oriented: New Zealand and Denmark. Lattimore presents the New Zealand policy towards competitiveness: Let the world market (and all its distortions) tell producers what they should produce without domestic distortions. This has worked well in many areas of New Zealand agriculture, and the country seems poised for major agricultural growth if world trade is liberalized. Walter-Jørgensen argues that the Danes do best in agricultural markets where government intervention is minimal. He believes that those Danish industries are competitive because of focused differentiation (focusing on a small number of differentiated products where world markets are large, product quality can be controlled easily, and productivity is important). It is interesting that Walter-Jørgensen is the only author to address wages in measuring competitiveness.

The final section of the book is a collection of papers by scholars given the responsibility to evaluate the lessons learned and to draw inferences of the value of competitiveness from the conference. The commentary of White provides an interesting analogy of competitiveness to healthfulness. He does not find the concept to be sufficiently rigorous to be very useful in economic analysis. Bullock, on the other hand, expresses the view that competitiveness, however defined, focuses attention on a new and different set of issues that he terms a new research paradigm. He argues that a number of important issues are not being addressed by agricultural economists at land-grant colleges and that a competitiveness-based paradigm identifies those issues. McCalla likens the two analytical approaches, marketing versus trade theory, presented at the conference to two distinct camps that don't talk to each other. He presents an interesting contrast to the very different ways that a traditional trade economist analyzes an issue with that of a "marketer" or an economist drawing from the industrial organization literature.

Our most important conclusion is that analysis and models must fit the problem or issue. Using nation-level concepts to reach conclusions at the sector or firm level, as is often the case, is clearly inappropriate, and using firm-level concepts to explain national competitiveness and to reach policy recommendations is just as inappropriate. The symposium offered trade economists and business strategists the opportunity to begin bridging the wide gulf between them.

Notes

1. We are greatly in the debt of Dr. Stanley Wilson, the former Executive Director of CAST, for his support.

PART TWO

Conceptual Foundations and Assessments from Business Strategies

2

Competitiveness: Definitions, Useful Concepts, and Issues

Philip C. Abbott and Maury E. Bredahl

Defining Competitiveness

A number of definitions of competitiveness, depending on the level of the analysis (unit of observation: nation, sector, or firm), the good analyzed, (commodity or differentiated product), and the intent of the analysis (policy prescription, sector productivity growth, export performance, etc.), have been proposed. The following illustrate this diversity:

> What we should mean by competitiveness, and thus the principal goal of our economic policy, is the ability to sustain, in a global economy, an acceptable growth in the real standard of living of the population with an acceptably fair distribution, while efficiently providing employment for substantially all who can and wish to work, and doing so without reducing the growth potential in the standard of living of future generations (Landau, 1992, p. 6).

> Seeking to explain "competitiveness" at the national level, then, is to answer the wrong question. What we must understand is the determinants of productivity and the rate of productivity growth. To find answers, we must focus not on the economy as a whole but on *specific industries and industry segments* (Porter, 1990, p. 6).

> The concept of comparative advantage describes the trading patterns that would occur as countries assess the relative costs of producing or trading in a world free of price distortions imposed by government policy. The real world, however, is replete with trade, agricultural and tax policies effectively overriding the determination of comparative advantage. . . . In brief, comparative advantage applies to a world of efficient, well-functioning, undistorted markets,

and competitiveness applies to the world as it actually is (Barkema, Drabenstott, and Tweeten, 1991, p. 254).

. . .being competitive is the. . .ability to deliver goods and services at the time, place and form sought by overseas buyers at prices as good as or better than those of other potential suppliers whilst earning at least opportunity cost returns on resources employed (Sharples and Milham, 1990, p. 1).

A competitive industry is one that possesses the sustained ability to profitably gain and maintain market share in domestic and/or foreign markets (Agriculture Canada, 1991, p. 3).

The first definition addresses factors broadly determining national competitiveness and policy evaluation in that setting; the second suggests that analysis at the national level does not contribute to the understanding of competitiveness; the third addresses the competitiveness of various sectors in agriculture and emphasizes the distorting role of agricultural policies; the fourth allows judgments of the competitiveness of particular agricultural sectors and introduces some notion of opportunity costs; the fifth is a pragmatic definition that describes sector performance in a meaningful way for business managers.

It is not surprising, then, that many different perceptions of competitiveness are championed in theoretical and applied literature and the popular press. Robert Reich, an early proponent of competitiveness, remarks that "rarely has a term in public discourse gone so directly from obscurity to meaninglessness without an intervening period of coherence" (Wall Street Journal, 1992). With that observation aside, the competitiveness discussion raises a number of issues that need to be addressed by agricultural economists and others interested in explaining the patterns of food trade, the impacts of trade liberalization, the role of government policies, and relative performance of sectors and firms.

We believe confusion on competitiveness stems from both *misunderstandings and inadequacies* of received theories, as applied to a wide range of trade policy questions. One goal of this chapter is to separate some of the misunderstandings from the inadequacies of theory. We also recognize that issues may be addressed at one level of theory, but results from that theory may beg questions addressed (or not) at a more profound and detailed (though vague and less general) level of analysis. For example, macroeconomic assessments of competitiveness beg the question: Why have factor productivity levels changed over time and across nations? But the highly aggregated models employed are not sufficiently complex to provide a complete answer. Most explanations offered go outside the bounds of the model, which first identified the troublesome productivity trends. Ricardo's model of comparative advantage begged the question: Why do costs differ across nations? The answer of Heckscher and Ohlin, that resource endowments account for these differences, is only one possible (or partial) explanation now recognized. *It is important at each stage of analysis to ask what our models are capable of examining and whether they are fully appropriate to the questions at hand.*

Some of the "mistakes" in the literature and in the popular press are examined first to see how wrong answers from received theory contribute to confusion on competitiveness. Useful concepts that have emerged from the trade literature that can aid in identifying causes of competitiveness are then reviewed. This review shows that, while many of the concepts explaining national competitive advantage (Porter, 1990) are found in the trade literature, this has not led to a general, unified theory that can accommodate all those concerns and that is useful for empirical analysis. A "continuum" of economic theories, ranging from the highly aggregate and elegant macroeconomic models to the disaggregated and detailed but fuzzy and sometimes *ad hoc* models of firm strategy, is presented to sort out what research issues may be addressed with what models and why each level of analysis is at times desirable. Based on these reviews, we propose some "right questions" for trade policy analysis and examine issues that must be addressed in analyzing competitiveness in food and agricultural trade.

Misunderstanding the Theory: Wrong Answers

Examination of some of the mistakes found in the literature and popular press in addressing competitiveness and trade is instructive in identifying some of the *misunderstandings of* received theory. Some answers purported to explain competitiveness simply violate economic theory, while others seek to apply theory to a question for which it is not appropriate. That is, the wrong theory is used to examine a question. It may be, as well, that at times economists and their theories don't take into account the concerns driving policy, as our discussion of outcomes measures for competitiveness assessment will demonstrate.

Production Costs Determine Location of Production

Basing judgment of competitiveness or comparative advantage on international comparisons of costs of production across nations is probably the most blatant violation of received theory in the literature. Ricardo taught us that comparative advantage means *relative, not absolute costs* must be compared. Thus, while the U.S. may be the least cost producer of corn, other countries may export corn if costs are low relative to their costs for other goods. Classic illustrations of the relevance of relative versus absolute costs abound in principles textbooks, yet we continue to see complaints in the press that other countries continue to export goods for which we are lower cost (or more efficient) producers. While we teach that relative, not absolute costs matter for trade, emotional arguments point to costs and inefficiencies across nations in decrying unfairness in observed trade patterns.

We also forget that Heckscher (1919) and Ohlin (1933) taught us that production costs are not fixed, but may vary with the level of output. Popular analysis often suggests that a product for which a country is a high cost producer should disappear from the production (and hence export) mix of that country. Modern economic analysis suggests that resource productivity and marginal cost can vary with the level of output so that as production decreases, a country will likely regain its competitiveness. The new "strategic" trade theories' recognition of economies of scale as an important factor further complicates this analysis.

Labor Costs Determine Trade Patterns

A second mistake is the focus on labor costs as the sole determinant of trade patterns. Ricardian theory identifies national differences in labor productivity as critical. Since there is only one immobile factor of production, generally called labor in Ricardian models, this is where those differences are attributed. In modern analysis, all components of value added, including returns to capital, taxes, and managerial skill, may reflect national differences leading to comparative advantage. Macroeconomic assessments have proposed total factor productivity in preference to labor costs as a measure of a nation's competitiveness. Both analyses of trade and of regional economic development cite a host of reasons why production may not move to where wages are lowest. Cost factors, such as taxes and utility costs, receive most attention, but a host of additional factors, including availability of educated human resources, infrastructure, local competition, quality of life, and other factors, also matter to choice of location by firms. These are reminiscent of explanations found in the regional development and business strategy literatures.

Capital Accumulation Determines Technical Progress

Another more subtle problem concerns the aggregate, homogeneous conception of capital, which emanates from the macroeconomic perspective. Accumulation of capital is seen as the means by which technical progress is implemented and the driving force behind changes upgrading sectors to maintain competitiveness. But the view of capital accumulation of a homogenous, productive, all-purpose machine for the production of the aggregate "good" is insufficient. This view of capital is simply inappropriate for analysis of competitiveness on a sectoral level and on the evolution of a product or sector. Capital should be perceived as a multi-dimensional asset, embodying numerous characteristics, which may be tied to a specific location, which may constrain certain location choices, or which is completely mobile. Capital specifications may also be related to the quality of product produced. Technical change may be cost reducing or may futun on enhancement of product quality. And our concept of investment needs to incorporate investment for upgrading and for keeping a firm at the leading edge of its sector. This upgrading may or may not require throwing away sunk investments—industry specifics matter.

In fact, economic theorists have struggled with the definition of capital and its relationship to technical change, as evidenced by the Cambridge-Cambridge debate several decades ago; by the various definitions of capital in the literature, such as the vintage capital model of Solow; by the sector-specific capital concept of Ricardo-Viner trade models; and by modern concern with endogenous technical change. The concerns driving these theories reflect the understanding that technical change and investment involves more than simply foregoing current consumption and the need to consider what form investment must take to make/keep an industry or firm viable.

From a practical perspective, macroeconomics has needed to use the simpler notions of capital in order to progress on other fronts. Indeed, the problem of aggregate savings is apparently one of the fundamentals behind aggregate productivity trends. But this aggregate perspective cannot explain the cycles of specific industries or products. Richer theories are needed to understand how competitiveness will develop and evolve.

Exchange Rates Will Balance Trade Accounts

Concern with the U.S. competitiveness "problem" has arisen largely due to the balance of payments difficulties, and yet it may not be competitiveness at all that has led to those problems. This view suggests that deterioration in the value of the dollar and sustained current account deficits are the consequence of U.S. products that have become unacceptable—hence not competitive—in international markets. But an understanding of international macroeconomics and finance, and recent theories on exchange rate determination, indicates that exchange rates are determined as much by foreign capital flows as by product or service flows. Macroeconomic (savings-investment) balances can be critical to current account and hence trade-balance outcomes, in spite of individual product competitiveness.

If the imbalances of the U.S. macroeconomy, due to a government spending in excess of its revenues and to a society unwilling to save for the future, leads to foreign capital inflows, then making U.S. products more attractive to foreign buyers will be a futile attempt to turn around a negative trade balance, which must persist to accommodate the necessary capital flows.

The willingness of foreign investors to move capital to the United States should be taken as a positive sign on the competitiveness of U.S. industries, not a negative one. But this perspective can be easily clouded, as small changes in macroeconomic policy, and especially interest rate differentials, can cause major swings in incentives to international capital movements. That volatility may be interpreted as a change in competitiveness, when it is the macroeconomic incentives that are driving the direction of foreign capital flows.

Minimizing Production Costs (Maximizing Consumption) Is the Right Objective

Differences in the assessment of competitiveness often reflect the differing objectives of analysts, who are coming from different perspectives. Inappropriate measures may be used when we do not clearly recognize goals of society or, more importantly, how those goals are reflected in the views of policymakers and in government actions.

A simple problem is that competitiveness assessments often focus on production costs as a criterion, when *value added* would be a better measure of the success or failure of a firm/product in trade. Value added counts the returns to labor, capital, and the government — the entities to whom the concerns of competitiveness are properly directed. This perspective reflects the broader concern reflected in the competitiveness debate, which goes beyond simply whether firms are successful, by also asking to whom the benefits are distributed.

Foreign ownership can affect the benefits from locating a firm in a country to that country or region, as can the sourcing of parts. Profits are of limited value from a nationalistic perspective if they accrue to foreign owners, though other secondary benefits may be derived from the activity. With inputs sourced from numerous locations, including foreign sources, the profitability of a firm and its contribution to the local area depends on not only the profit it generates, but also broadly on the employment and income it generates, and on who receives that income. In this respect, measures of value added are more illuminating than profitability, per se.

Some more recent assessments of competitiveness have advanced on this concern with an appropriate measure of benefits, recognizing that in our economic analyses, the ultimate objective is *consumer utility maximization*. That is, the goal of the economy is to meet the wants and needs of society, captured in an economist's simple view of the world by the level and pattern of consumption. Profit generation by firms is good only insofar as it is a means to that end.

It must be recognized, as well, that economists' objectives may not correspond well with those of government policymakers and of other social interest groups. The best example is the concern by many groups with the generation of quality employment opportunities, rather than the generation of consumption as the target of policies, or the yardstick to measure competitiveness. This concern with *job creation*, beyond production and profits, is seen in the policy debates of many nations, at varying stages of economic development. It probably reflects the perspective that humans derive worth from what they produce in addition to what they consume. And work is more than a means to generate income, but also an important aspect of one's life and hence an individual's assessment of their self worth. By ignoring these perspectives in our evaluation of competitiveness, economists can easily become irrelevant to the policy debate. Simple models have trouble incorporating these multiple and at times fuzzy objectives, however.

Useful Concepts from Trade Theory

Existing trade theory can carry analysis of competitiveness farther than is often assumed. Factor endowments remain an important, if incomplete, explanation of trade and production patterns. Many of the concepts found in modern debate on competitiveness (e.g., Porter, 1990) find their roots in old trade literature, especially from post-Leontief paradox discussions. Leontief tested empirically in 1956 the standard approach to theoretical trade analysis, the Heckscher-Ohlin (H-O) model, and found it inadequate to explain observed trade flows. This led to a flurry of activity incorporating new concepts into the explanation of trade. But several decades later, problems with the standard paradigm remained, and the explanations that emerged from the Leontief paradox had become caveats for theoretical and modeling exercises, rather than leading to a new, unified theory for empirical analysis. A later development was the marriage of trade theory with industrial organization theory, which demonstrates that in individual industries, institutions, infrastructure, and technical characteristics matter. This literature teaches us that a unified, general theory to address competitiveness may be unattainable — since industry specifics matter. But a number of useful concepts, reviewed below, may be gleaned from this literature.

Addressing Inadequacies of Received Theory

Among the first explanations of the failure of the H-O model to explain observed U.S. trade patterns was the recognition of the possibility that *demand conditions* varied across nations. Contrary to the assumptions of H-O, tastes and preferences could vary across nations. Linder (1967) explored the possibility that even if utility functions were identical, varying income levels would lead to differing consumption patterns across countries. Preferences are simply not homothetic, as Engel's law continues to show us. Porter (1990) argues that product differentiation leads to a need to recognize country-specific preferences.

While these explanations would not account for the entire failure of H-O theory, modern thought on national competitive advantage clearly draws on these old ideas. Yet trade theorists resisted, because homothetic preferences permit robust theorems to be derived, while differing demand patterns (and utility functions) can turn around otherwise useful conclusions. This does not mean in our search for why firms succeed or fail that differing cultures can be ignored, while theoretical developments will continue to find it useful to put these concerns to the side for a moment.

The role of *natural resources* in explaining production and trade patterns was raised by Vanek (1959) and Kenen (1965) as another potentially erroneous assumption of the H-O model. Some progress was made in explaining certain trade and production patterns following this reasoning. Especially in agriculture, the production of certain goods may be tied to a given resource base. Specific crops require specific land and climatic conditions.

This concern raised the broader question of whether two factors was sufficient to explain observed trade. Among the many factors introduced were alternative and disaggregated versions of capital. Indeed, the concern with an appropriate *definition of capital* was recognized soon after the Leontief paradox. This also raised concern over effects due to technical change. One of the first innovations, and among the most important explanations of the failures of the H-O model, was to recognize the possibility that *human capital* endowments could matter separately from physical capital. Another direction was the relative requirements for *research and development* inputs by industry. Vernon (1966) took these questions further by considering the *cycles of products*, which would be produced in the country of discovery initially, but whose locus of production could change as the product matured and technical information spread.

These product-cycle and technical-change questions naturally led to concerns with the *dynamics* of trade and hence "dynamic" comparative advantage. The stock of capital in a country may be fixed, and even sector-specific in the short run, but in the longer run the constraint is savings, not physical capital. The possibility of foreign capital inflows weakens this constraint, since otherwise profitable investment opportunities will not be limited by domestic savings alone. And the form or nature of investment may depend on a product cycle, the extent of sunk costs, and the possibility of upgrading and hence flexibility of investments. Dynamic comparative advantage asks what are the determinants of and constraints on technical change and in turn on how that is embodied in and implemented through investment strategies. Theory is not tight on this question, and empirical explanations of investment patterns at an aggregate level continue to flounder.

Another important line in trade analysis, raising again the question of appropriate benefit measures, is the *theory of second best*. Observed trade patterns are the consequences of distortions in place, some of which may be irrational, but others of which may be introduced to achieve desired noneconomic objectives. And from differing perspectives, what constitutes an appropriate noneconomic objective and what constitutes irrational policy may differ. One branch of this theory recognizes that some distortions are introduced to achieve noneconomic objectives, and that other distortions are immutable so that policy is needed to address them (Bhagwati, 1971). The consequence is trade patterns that differ from cost-based comparative advantage.

"Strategic" Trade Theory

While the Leontief paradox led to a plethora of constructive new ideas on what explains trade, most explanations that emerged represented minor deviations from necessary assumptions or extensions into a world of multiple dimensions. These explanations remained unsatisfactory, and a marriage of trade theory with modern industrial organization theory arose in order to introduce several new concepts into the analysis.

A few key assumptions lie at the heart of the new "strategic" trade theory. *Economies of scale* are seen as important for certain sectors. *Product differentiation* is also critical, as well as the critique raised above that products must meet the specifications of consumers, possibly for segmented markets. Not only do consumer tastes vary across countries but also within countries. Coupled with economies of scale, cross-hauling of the same general product embodying specific characteristics is not at all an unreasonable outcome.

While scale economies and product differentiation correspond to the characteristics of production and demand, *market institutions* may also matter, according to this literature. Imperfect competition and institutional arrangements can cause optimal trade patterns and policy to differ from conventional results. Another part of that literature has also found that firms may price to market (Krugman, 1987), since trade barriers, imperfect competition, product differentiation, and scale economies can raise entry barriers and permit markets to operate independently. While transportation costs alone may limit opportunities for arbitrage, these institutional market characteristics and the realities of doing business can at least slow the movement toward equilibrium and allow price differentials to persist.

More recent literature has taken on the questions of incentives to and directions in *technical change*. The same market institutions and economic assumptions that can alter trade outcomes are seen to affect endogenously the evolution of technical change. In agricultural development, Hayami and Ruttan (1971) argued several decades ago that agricultural technical change has followed different paths according to local economic conditions. Their explanation was based on factor endowments and so was seen as related to H-O trade explanations, but national differences in factor prices flew in the face of the factor price equalization theorem and its implications for national technological choice. This modern recognition of pricing to market, imperfect competition, and endogenous technical change offers a resolution to that dilemma.

While general policy prescriptions have been scarce from this literature, and often the recommendation is that policy intervention is not justified, some important lessons do emerge. Specific economic issues that impinge on competitiveness and trade patterns have been identified and explanations of some observed anomalies were forthcoming. More importantly, we learned that market institutions and industry specifics matter, both to trade and production outcomes and to policy.

A Continuum of Useful Concepts

Trade theory is but one of the subdisciplines in economics concerned with issues related to competitiveness. We have found it useful to establish a "continuum" of economic theories and subdisciplines, ranging from the elegant but highly aggregate macroeconomic analysis to the detailed case study approach of firm strategists. As one moves from one level to the next, theories and concepts become at the same time more detailed and complex, yet less precise and elegant. Each level is not a discrete step, however, as these subject areas will of necessity overlap.

The strict assumptions of the macroeconomic and trade theory perspectives have permitted derivation of robust general theorems (within the limits of the assumptions imposed), while approaches closer to practical reality, where the limiting assumptions are unacceptable, have not led to tight theories or general models. But the more general theories offer only incomplete answers to competitiveness questions and beg many of the subsequent questions the more *ad hoc* approaches seek to address. Each level of analysis is at times appropriate, but leads to problems when it is pushed beyond the limits imposed by its assumptions. While macroeconomic theory can point to investment trends limited by low saving as a source of declining total factor productivity, it is ill-suited to examination of industry specific policies. The strengths and limitations of macroeconomic, trade, microeconomic and firm strategy approaches to competitiveness analysis are briefly reviewed below.

Macroeconomics. A macroeconomic perspective is useful in recognizing the aggregate trends and underlying problems of an economy. Its clarity and the elegance of its theorems derive from its simplicity. The world is seen in terms of one homogeneous good, two productive factors, and relatively few, similar economic agents. In the analysis of competitiveness, the macro perspective is where the problem of declining total factor productivity is first seen. Combined with international finance, the various problems of macroeconomic balances have also been highlighted. The fundamental balances are between investment and savings, from both domestic and foreign sources. This approach clearly demonstrates the negative consequences of declining propensities to save, of rising government deficits, of trade and payments imbalances, and of foreign capital flows.

The problem with this approach is that it is just too aggregate to reveal all the fundamental causes of the trends it uncovers. We know that factor productivity was not growing as fast in the United States as elsewhere, but is that a normal consequence of our stage of economic development or a symptom of some economic malaise? The macroeconomic perspective has shed light on some causes of this trend, but most of the explanations offered fall outside the boundaries of the macroeconomic models that revealed those problems.

We must at least move to a two-good world, permitting exports and imports, to understand how different sectors fare and why trade arises as it does. Most explanations of trade patterns, however, go into far greater detail on the microeconomics of specific sectors and in changes in the composition of the economy and its effects.

The macroeconomic approach was useful in setting the trends and highlighting problems, but the questions it begs must be answered at a more refined and disaggregated level of analysis. Issues of sectoral performance or the costs and usefulness of various subsidies simply cannot be answered within macroeconomic models. Its strength lies in the insights and overall perspective it offers. Its weakness is that it only raises but does not answer most questions on how competitiveness can be fostered and maintained.

Trade Theory. Trade theory addresses competitiveness issues at two distinct levels. The bread and butter framework of trade theorists — the two good, two factor Heckscher-Ohlin-Samuelson model — is only an extension of the macroeconomic approach. This model has been extended at a theoretical level to many goods and factors, with a loss of elegance and with far greater limitations on the theorems that fall out. Basics such as factor price equalization quickly fall apart in this multidimensional world, although certain other basic principles survive. Factor endowments at a disaggregated level remain important explanations of observed trade patterns. Computable general equilibrium models have permitted the empirical implementation of this disaggregated perspective.

The trade theory presentation in this book goes back to several of the arguments arising from the macroeconomic perspective. The most useful perspective concerns the marriage between trade, macroeconomics, and international finance. Trade theory also raises certain questions not apparent from the macroeconomic framework. What should be imported/exported? When are trade policies appropriate and which are counterproductive? Is there a special role for trade policy apart from sectoral or macroeconomic policy? Trade theory can examine these questions, considering intersectoral linkages in a general equilibrium context, a perspective that is subsequently lost in the more microeconomic approaches.

The most useful insight arising from the trade literature that addresses the evolution of competitiveness and the appropriate sectoral policies comes from the empirical failure of standard models, not their successes. The Leontief paradox generated many of the ideas found in Porter's national competitive advantage. The new strategic trade theory, which arose out of continued questioning of standard approaches, further illustrated the range of factors related to the success or failure of specific industries, and the role of trade policy in influencing competitiveness. Factors identified include demand conditions, product differentiation, natural resources, technical change, human capital, product cycles, scale economies, and imperfectly competitive market institutions, among others. Sector-specific market institutions are emphasized, and the lessons derived are seldom general in application to a wide range of industries. The lessons from these areas serve as the caveats for theorists and modelers but are the grist for those seeking to explain competitiveness in specific cases.

While trade theory is the framework necessary for cross-sectional, multi-product comparisons, it also raises microeconomic questions that may or may not be answerable at as general a level as required by trade models. The strength of the trade models is that certain issues can be seen in an aggregate perspective. As for the macroeconomic perspective, ultimately the success or failure of a sector, product, or firm depends on microeconomic detail. *We simply doubt that a few key factors can be identified that explain the pattern of trade at an aggregate level, or whether a good should be imported or exported.* The current answer can well be that trade flows will go in both directions. The role of trade policy relative to domestic policy can be clarified, especially based on the theory of second best, which corresponds more closely to the world in which real decisions must be taken. Institutional details matter, so policy prescriptions will differ by sector.

Microeconomics/Business Strategy. A microeconomic and/or business strategy approach is the final level of our continuum, representing the most disaggregated and detailed level of analysis to questions of competitiveness. Microeconomics relies on profit and utility maximization paradigms. Business strategists look at an even greater set of issues, freeing themselves of the theoretical constraints of microeconomic theory (maximization principles) at times. While this approach will find difficulty in examining intersectoral links and cross-product comparisons, it is the most likely to identify appropriate policy for a given sector or the likely success or failure of that sector. The detailed, sector-specific information generated can be fed back into the trade and macro models, to some extent, to look at the broader policy issues. But one of the empirical lessons that we believe has emerged is that both constraints and advantages are seldom absolute. Production of certain goods can persist in both high and low wage nations (e.g., computers) for a variety of reasons, due to such factors as market segmentation and the differing level of service and reliability demanded by those different markets. Countries with an inherent advantage can also stagnate and lose an industry if technical change passes them by. Generalizations on what is sufficient to guarantee competitiveness is apparently quite elusive, and complex explanations arise as to specific successes or failures. Case studies are illuminating in retrospect and can highlight important issues but will not lead to generalizable answers.

Right Questions

The above discussion of trade theory and its place in the context of other approaches to economic analysis suggest some more appropriate questions for analysis of competitiveness and trade policy. The classic *text book questions* remain at the heart of our work, however. We fundamentally wish to explain what does (and should) determine the observed patterns of trade and production across nations. This is exactly the question posed in the post-Leontief paradox debate and remains central to our concerns.

What Determines Investment? Taken to a more practical, applied level, trade questions need to be reformulated to ask what are optimal patterns of investment by firms/nations and what policies are appropriate to the realization of optimal investment patterns. The evolution of firms, the dynamics of production, and consequently the observed patterns of trade are the results of successful (or failing) investment strategies. Dynamics and timing clearly matter.

What Determines Firm Success? Business strategy analysis takes the question of optimal investment to a more specific, detailed level, asking what factors influence the success or failure of firms. The definitions of competitiveness by business school types reflect this concern, measuring the success of firms in terms they see as relevant, leading to the emphasis on market share as well as profitability. This approach has introduced some new concepts that take us beyond the trade literature — firm strategy, marketing and distribution channels, first mover advantage, the role of innovation dynamics, and clusters and rivalry.

What Government Policies Are Optimal? Central to much of our literature on competitiveness and trade is the question: What, if anything, should governments do about the competitiveness of domestic firms? In trade theory jargon, this asks: What is the optimal mix of policies, and what role should trade policy, per se, play in that mix? Raising that question poses the same dilemmas faced in the debate on the theory of second best. Whose welfare is at stake? What distortions are the consequence of misguided government policies, and which are characteristics of industry structure? Which distortions are subject to change, and which are immutable and so define the environment within which policy must be set? Which policy instruments are politically acceptable and which are not?

Some have drawn a distinction between competitiveness and comparative advantage based on the distortions introduced by policy (Monke and Pearson, 1989). Comparative advantage is the ideal we seek in a undistorted world, whereas competitiveness seeks to address trade policy questions within the distorted world in which we live, ranking options given those distortions. If that distinction is accepted, comparative advantage is relegated to a sterile theoretical concept of little practical value. Competitiveness simply replaces it.

The new "strategic" trade theory has shown that industry structural details matter in setting optimal policy and that the basis for intervention is often weak, even in an imperfectly competitive setting. Porter's analysis sees only a very limited role for the state, to ensure the appropriate regulatory environment and infrastructure within which firms realize their competitive advantage. Few approaches see subsidies as being an optimal or even second best government strategy. The role of government is generally seen to ensure a competitive economic environment, to provide the public goods and infrastructure that private firms will not offer, and to set regulations to ensure the health and safety of its population and environment. If subsidies are pursued, inevitably some industries are favored over others, and rent-seeking behavior rather than optimal resource allocation will dictate policy. While government can make firms fail, it cannot force firms to succeed.

Determinants of Competitiveness

Porter's "National Diamond" is generally used in the context of case studies of competitiveness to assess the prospects of an industry, product, or economic activity. That approach represents one useful grouping of the concepts appropriate to competitiveness and trade analysis, although it may suggest his theory is simpler than it is in fact. We propose below a slightly more detailed categorization of factors determining competitiveness, which follows somewhat more closely the approach used by trade economists.

Factor Endowments and Natural Resources. The standard trade paradigm pointed to factor endowments as the key to comparative advantage. Failure of that model to predict observed trade patterns led to an understanding that disaggregated factors of production, and especially natural resources, can be among important factors conditioning competitiveness. Natural resource advantages are of particular importance to agricultural commodities, since soil, climate, and other natural conditions can determine where crops may be successfully grown.

Technology. Country specific advantages or disadvantages determined by technology serves as the fundamental basis for competitiveness assessment. Technical change may be both *cost-reducing* and/or *quality-enhancing*. Some aspects of technology may be highly mobile across nations, while especially for resource-based enterprises, national characteristics (climate, land quality) can be the determining factor. Often technologies pose constraints that are more important for some countries than others (e.g., waste disposal from animal production). For a country to have a competitive advantage, it must have access to appropriate technology, which may require investment (public?) in research and development.

Investments. Investments are the means by which technical change and industry evolution are accomplished. Market and technical factors both impact investment strategies. Some enterprises will offer opportunities for technical upgrading, while sunk costs have at times trapped nations in declining industries. Proper investment strategy combines the technical dimension with marketing and business strategy.

Human Capital. Availability of human resources is critical to competitiveness of specific products or sectors. Few enterprises require truly unskilled labor. Reich has pointed out that less and less of product value is the manufacturing cost per se, based on generic labor input, and more and more is derived from services whereby products are tailored to customers' specific needs. Expertise is critical to an enterprise, but it need not be offered from the same nation that produces the final good sold.

Managerial Expertise. An important component of the human capital dimension is management expertise. Case studies point to firm failures, when all the signals seemed to point to success, because of mistakes made by management. Individual firms may not always pursue the optimal, long-run profit maximizing strategy, and the quality of management is likely a critical factor.

Product Characteristics. As the value of an export increases, and as we move from commodity and natural resource based exports to higher value added exports, product characteristics become increasingly important. We have seen that tastes and preferences can vary across or within nations, and successful business strategies can be designed around serving market niches in addition to broad market demands. Other *non-price factors*, such as reliability, maintenance, and service, can be important components of product characteristics to both processors and final consumers.

Firm Strategy and Industry Structure. Market institutional characteristics can also affect competitiveness and can vary substantially across products. As seen for product characteristics, there may be several successful firm strategies. Cost leadership is one option, whereas serving market niches may not demand the lowest production cost. In many instances the nature of competition in an industry can affect the outcome of strategies or policies intended to impact on those strategies.

Input Supply. Few enterprises are completely vertically integrated. Relationships between producers and their input suppliers can be critical to the success or failure of a firm or product. But firms may source inputs from other nations, while sometimes the availability of local suppliers can offer a substantial advantage to an enterprise. National benefits to firm location decisions may be substantially derived from benefits accruing to these related industries.

Marketing and Distribution Channels. The system to market products, and especially to penetrate export markets, can be crucial to the success of firms. The internal transportation network of the U.S. is seen as giving an important advantage in export competitiveness for grains, while European ties to North Africa and the Middle East may give an advantage in their value added exports to that region.

Infrastructure and Externalities. Government is responsible for the necessary infrastructure, including public works, utility regulation, education, and other public goods. Often the determining factor is external benefits that accrue not to a single enterprise, but rather broadly to many possible ventures.

Regulatory Environment. The government sets the rules of the game under which firms must proceed, and these can be exceedingly specific and complex, constraining the decisions and opportunities of firms. While regulations are clearly necessary, for example to protect the health and safety of the population, differing perspectives on what constitute appropriate regulation, or different values based on such outcomes as environmental quality, can cause these rules to differ across nations. Governments must decide which regulations are in the public interest, and what actions (if any) are necessary or desirable to foster particular enterprises.

Trade Policy. Trade policy is a special case of the set of regulations imposed on a firm, relating to products crossing borders. In agriculture we have found that it is the set of domestic policies (or at least domestic objectives), more so than trade policies, that set the environment within which firms compete. A critical yet unanswered question is: When is trade policy per se an appropriate strategy for government? Few believe that protectionism is often in the public interest. More often, policies directed at distortions rather than via trade are to be preferred.

Conceptual Framework for Agricultural Competitiveness

We found it necessary to condition our identification of factors affecting competitiveness and the application of trade theory to international food markets on selected characteristics of the several "levels" of food and agricultural production and processing. Competitiveness in international food markets, or in the domestic market with imports, is determined by decidedly different factors and measured by different instruments depending on the type of product produced.

Our starting point is the seminal 1960 article by Breimyer, in which he introduced the notion of "three economies of agriculture" — production of primary products, conversion of feedstuffs into animal products, and marketing of food products. This conceptual framework was used to analyze, among other things, the relationship of ownership and management of resources, implications of the relationship of the final consumption good to the primary product, and price trends.

It proved useful here to define four economies of agriculture that could serve as the basis for identifying an appropriate definition of competitiveness and the corresponding appropriate analytical framework and for identifying factors affecting the location of production and the patterns of trade, etc.

Like Breimyer's, the taxonomy developed below is *defined in terms of the characteristics of economic activities* (e.g., level of processing). In particular, it is based on the degree of potential substitution among traded and nontraded inputs, the linkage of primary production to end-use characteristics of final consumption goods, the relative importance of product versus process technology, and the resulting value added in the economic activity. The four economies of agriculture are:

- production of an *undifferentiated primary commodity* with little or no linkage between production and end-use characteristics in final consumption;
- production of *differentiated primary products* where a linkage likely exists between production and processing and end-use characteristics in final consumption;
- conversion of primary products and commodities into *semiprocessed products*; and
- conversion of primary and semiprocessed products into processed *consumption-ready products*.

Selected issues in competitiveness in international food markets identified earlier are presented within this context in Table 2.1. We attempt to rank the importance of each of the issues for each of the four economies.

Each column identifies the important determinants in the competitiveness of each of the types of economic activity. For example, natural resource advantages, cost-reducing technology, infrastructure, and trade/domestic policies were the issues judged most important to competitiveness in global commodity markets. Quality enhancing technology, human capital, firm strategy, industry structure, and non-price factors were judged to be of lesser importance. For semiprocessed products, cost-reducing technology and government policies were judged to be of lesser importance. Issues of importance for that sector and for the consumption-ready product sector included firm strategy, industry structure, and quality of products. These issues could be of paramount importance where the end-use characteristics of the final consumer products depended on production processes at the primary production level.

Evaluating each element of Table 2.1 without regard to interactions across elements is not appropriate. Commodities closely linked to a natural resource can become differentiated products (identity-preserved cereals, for example) in which case the second column identifies the important issues in competitiveness. Firm strategy and industry structure cannot evolve independent of the type of economic activity involved.

TABLE 2.1 Importance of Selected Determinants of Competitiveness in the Four Economies of Agriculture

Determinants of Competitiveness	Production, Assembly, Transformation (Processing) and Final Distribution of:			
	Undifferentiated Primary Commodities	Differentiated Primary Products	Semi-processed Products	Consumption-Ready Products
Natural Resource Advantage, Factor Endowments	Generally critical, but the mobility of technology is likely reducing its importance.		Little importance, but varies with mobility of primary outputs.	Little importance, but varies with mobility of primary and semi-processed products.
Cost-Reducing Technology	Mandatory, but technology is increasingly mobile.		Some importance, but product differentiation requires certain characteristics be reflected in production practices, technology generally mobile.	
Human Capital and Managerial Expertise	Some importance; skills in application of production technology important, many people involved.		Great importance; skills are critical, especially in organization and coordination of activities, with fewer people involved.	
Quality Enhancing Technology	Some importance: quality, transportation, etc.	Some importance: Quality/ product form	Great importance; end-use characteristics most important	
Infrastructure	Important to cost competitiveness.		Important to cost competitiveness and product differentiation; and to innovation.	
Product Characteristics and Non-price Factors	Some importance: grades and standards provide information.	Moderate importance: product differentiation possible through quality differences.	Great importance: degree of product differentiation and other activities determine the amount of value added.	

Firm Strategy	Minimum cost is only feasible strategy.	Some importance: cost and differentiation are possible strategies.	Great importance: cost leadership and product differentiation, or a combination may be pursued.
Industry Structure Input Supply, Marketing and Distribution	Some importance: markets provide vertical coordination.	Importance varies depending on economies of scale in economic activities other than production. Markets or hierarchies link primary product production. Often accomplished by single firms. Importance of end-use characteristics at farm level varies, and influences the vertical coordination of markets.	
Infrastructure	Important to cost competitiveness.		Important to cost competitiveness and product differentiation; and to innovation.
Regulatory Environment and Trade Policies	May determine trade patterns.	Importance varies; policies greatly influence competitiveness and trade patterns. But, often the policy impacts are indirect. Technical barriers matter most.	

Issues in Agricultural Competitiveness

The most pervasive concept of competitiveness practiced by economists and agricultural economists is *price competitiveness*. So, national agencies (the Bureau of Labor Statistics, for example) and international organizations (Organization for Economic Cooperation and Development, for example) compute and publish measures of productivity and unit costs (Hickman, 1992). Agricultural economists evaluate and compare trends in yields and output unit costs and, on rare occasions, total factor productivity or other more comprehensive measures.

The preoccupation of agricultural economists with evaluating price competitiveness is illustrated by the great importance placed on the elasticity of export demand for U.S. agricultural products, the proliferation of net trade models constructed from estimated and synthesized elasticities with little or no attention paid to market structure and other competitiveness issues, and the reliance on price transmission elasticities to capture the complex impacts of trade and domestic agricultural policies on market structure, technological growth, and technical efficiency. But, as the earlier discussion illustrates, a host of issues should be consider simultaneously, not independently, in evaluating competitiveness and in prescribing policies to enhance it.

Non-Price Factors

The importance of non-price factors was illustrated in the above discussion, but our concept of it must be expanded to appreciate the range of issues that must be analyzed. Piercy (1982, p. 115), for example, questions the typical level of analysis: "The problem in this present (1982) context is that the bulk of this effort has been at the macro-economic level, concerned with comparing international competitiveness between whole countries." He concludes that ". . .competitiveness depends on price — in the full sense of cost to the customer, including discounts, currency effects, insurance, freight, credit and the like — and the many non-price factors connected with the product and how it is marketed — which have value to the customer and which influence his decisions" (p. 112).

Piercy identifies two broad categories of non-price factors: product factors — these are factors that relate to the product itself and include but are not limited to design, quality, performance, reliability in use, packaging, and physical presentation; and services — explicit and implicit service provisions and content including advice and assistance in making product choices, user education, training, after-sales service, and maintenance, and implicit services such as personal visits to the customer, speed in price quotation, speed and reliability of delivery, stockholding for fast supply of parts and spares, and technical advice. Rarely are this range of factors considered in the evaluation of the competitiveness in international food markets.

Production assembly, transformation, and marketing must be vertically coordinated to provide products with desired end-use characteristics in international markets. The emergence of hierarchies (such as contract production), in place of markets, might be viewed with a good deal less alarm if this need for vertical coordination were recognized by agricultural special-interest groups.

Productivity, Growth, and Technology

All definitions of competitiveness admit that productivity growth, capital formation, and technology are prime determinants of long-run competitiveness. However, our knowledge of the causal factors leading to differences in each of these economic outcomes is very limited. We have numerous cross-country comparisons at a point in time, and some time series analysis, but few studies that attempt to relate superior levels of productivity growth and technological innovation to government policies, firm strategies, or industry structure.

Of considerable importance for food and agricultural sectors is the interplay of firm strategy and technological change. The U.S. meat-packing industry is often accused of a management fixation on reducing unit costs, for example. These firms follow a cost-minimization strategy in contrast to a product differentiation or combination strategy. So, they encourage production of fat hogs and in the process reduce the competitiveness of the United States in international markets. That strategy places severe restrictions on technological innovation and productivity growth in the production of pork at the farm level. It also creates products that are not competitive in foreign markets.

In a broader sense, technology that is mobile across national borders does not provide a competitive advantage to the innovating nation. Technology that "sticks" is the desired alternative.

Foreign Direct Investment

Expanding the research agenda of agricultural economists to include direct foreign investment is important for two reasons. First, we need to understand the relationship among direct foreign investment, value added processes, and the balance of trade. United States multinationals seem to have an unusually high propensity to invest in foreign countries rather than to produce and export from the United States. Understanding that phenomenon is critical, and much remains to be explained. We also need to understand it to evaluate the criticism in the popular press that direct foreign investment exports American jobs, and so it reduces the potential contribution to the balance of trade. Does it reflect the lack of competitiveness of American workers or the negative impacts of domestic investment policy and other policies? Or does it reflect trade restrictions and export subsidies in foreign countries?

The second issue is the role that foreign direct investment will play as free trade areas grow in importance. Will high-labor content food production flee to low-wage areas? What government policies might positively affect that outcome? Notably absent in analyses by agricultural economists of the North American Free Trade Agreement is consideration of international capital flows.

The location of production, and hence the pattern of trade, may be affected more by foreign direct investment than any other factor in the 1990s, and the pattern of trade predicted by factor endowments may be even less valuable. For example, the future location of pork production and trade will likely not reflect the ability to produce a surplus of cereals, oilseeds, and other feedstuffs. The reason is the international movement of capital and the mobility of technology. A large U.S. multinational is the largest processor of pork in Taiwan. Danish interests are attempting to replicate the Danish pork system in Australia and Mexico. Taiwanese interests, in response to pressures to reduce environmental pollution, are evaluating investment in China and, of all places, in British Columbia. There does not appear to be an immobile factor that is important enough to determine the global location of pork production.

Trade and Domestic Policies

Agricultural policy analysts have often assumed perfectly competitive domestic and international market structures, which implies homogenous products across competing export suppliers. These assumptions justify use of aggregate price responses (and related price elasticities) to evaluate domestic policy alternatives, to estimate the impacts of trade liberalization and of other economic events such as exchange rate changes.

These analyses have served well in the evaluation of aggregate policies, but recent studies of the impacts of trade policies and international comparisons of competitiveness suggest the need for disaggregated, and often product-specific, analysis. Hayes, Green, Jensen and Erbach (1991) illustrated the effect of the level of aggregation and adjustments for product differences on predictions of the international competitiveness of beef production and processing in several countries. They found that predictions of competitiveness varied with the market level at which the comparison was made. Rarely did farm-level price comparisons indicate the direction of competitiveness found at higher levels of processing. Reca and Abbott (1992) found that Argentine beef exporters, logically, filled their quotas for the EC beef market with mostly high-value cuts. Import unit values captured some of the quota rents and so would serve as a very imperfect indicator of competitiveness in the European market. Forsythe, Bredahl, Abbott, and Reca (1991) found that the disease status of beef-exporting countries dominated beef trade patterns. The world is effectively divided into two areas, disease free and infected, which determines the pattern of trade.

Bredahl, Sliffe, Reed, and Tvedt (1992) found that until recently the administration of the Japanese variable levy system for pork disadvantaged those exporters that could not provide a range of products meeting Japanese market requirements. The reason was that a combination of high- and low-priced products was needed to minimize the variable levy. For that reason, the United States lost market share to both Taiwan and Denmark after the mid-1970s. They also found that the liberalization of the Japanese pork market since 1986, which reduced the need to minimize the variable levy, had significantly increased the market share of Taiwan at the expense of Denmark. The reason was that the Danes could no longer export their pork loins, with lower quality characteristics, in combination with bellies and hams. The Taiwanese were found to be illegally exporting a significant part of their exports, evading the variable levy completely. The impact of market liberalization was found to involve a good deal more than the response of Japanese producers and consumers to price change.

The liberalization of the Japanese beef market also illustrates the importance of factors other than price in evaluating competitiveness. Despite numerous studies, utilizing concepts ranging from simple elasticity models to complex consumer demand systems and to political economy, few, if any, agricultural economists can claim to have predicted the outcome of the liberalization of the Japanese beef market. Overlooked in the analysis, the factors that turned out to be decisive were new entrants into the market and the emergence of new marketing channels and contractual relationships. The United States has captured the lion's share of the benefit from the liberalization by directly contracting with supermarkets for the sale of fresh (unfrozen) beef cuts. The large consumer market in the United States allows processors to ship only that portion of the beef carcass desired in the Japanese market and to supply it on a timely basis. This new marketing channel cannot be exploited by Australia or New Zealand; their advantage in longer shelf-life has been neutralized. With freer trade in other products, will the large consumer market of the United States provide a competitive advantage over other exporters?

These examples illustrate that analysis of price competitiveness simply was not sufficient to predict the outcomes of trade policies. They illustrate the value of analysis conducted in a disaggregated framework that considers trade by product and by source. Product analysis, contrasted with price or commodity analysis, is needed to provide an adequate picture of policy impacts and the likely outcomes of trade liberalization.

Conclusions

Competitiveness assessment requires that we go well beyond the limits of traditional trade theory to determine the pattern of trade and how it is influenced by firm strategy and government intervention. A host of factors beyond price competitiveness were identified, and an appropriate role for each of several levels of economic analysis was presented. We see the search for a single new paradigm to replace traditional trade theory as futile, but that was already well recognized in assessments of the policy significance of the new "strategic" trade theory. An important lesson of that literature is that industry specifics matter. There is then a need for the detailed case studies of business strategists to complement the theoretically rigorous but highly aggregated macroeconomic and trade/general equilibrium approaches.

Returning briefly to Table 2.1, we also believe that the profession has overemphasized the importance of commodities at the expense of the processing, product differentiation, and firm strategy dimensions of creating competitive exports. Competitiveness assessments of agricultural economists have also focussed on natural resource endowments and cost-reducing technology, which may be the critical factor for commodities, but for differentiated and processed products, quality, service and other non-price factors demand more attention. This book should represent a first step in laying out the research agenda needed to determine factors critical to competitiveness in international food markets for all types of food products.

References

Agriculture Canada. Task Force on Competitiveness in the Agri-Food Industry. "Growing Together: Report to Ministers of Agriculture." Agriculture Canada, Ottawa, June 1991.

Barkema, A., M. Drabenstott, and L. Tweeten. "The Competitiveness of U.S. Agriculture in the 1990s in Agricultural Policies." In *The 1990s in Agricultural Policies in the New Decade*, Edited by Kristen Allen. Washington, D.C.: Resources for the Future, National Planning Association, 1991.

Bhagwati, J.N. "The Generalized Theory of Distortions in Welfare." In *Trade Balance of Payments and Growth: Papers in International Economics in Honor of Charles P. Kindleberger*, Edited by J.N. Bhagwati, R.W. Jones, R.A. Mundel, and J. Vanek. Amsterdam: North Holland, 1971.

Bredahl, M.E., K.J.Sliffe, M. Reed, and D. Tvedt. "Product Effects of Trade Restrictions in the Japanese Pork Market." Working Paper 92-1, Center for International Trade Expansion, Missouri University, January 1992.

Breimyer, H. "The Three Economies of Agriculture." *Journal of Farm Economics* 44 (August 1962): 679-99.

Caves, R.E. and R.W. Jones. *World Trade and Payments: An Introduction.* 4th ed. Boston: Little Brown & Co.,1985.

"Competitiveness is a Big Word in D.C.; Just Ask the V.P." *Wall Street Journal* July 1, 1992: 1.

Ethier, W. *Modern International Economics.* New York: Norton, 1983.

Forsythe Jr., K., M. Bredahl, P. Abbott and A. Reca. "Impact of Animal Health Regulations on Trade in Livestock and Livestock Products," *Impacts of the Unification of the EC Internal Market*, Economic Research Service (USDA), October, 1991.

Hayami, Y. and V.W. Rutan. *Agricultural Development: International Perspective.* Baltimore: Johns Hopkins Press,1971.

Hayes, D. J., J. R. Green, H. H. Jensen, and A. Erbach. "Measuring International Competitiveness in the Beef Sector." *Agribusiness: An International Journal*, Vol 7 (1991): 357-374.

Heckscher, E. R. "The Effects of Foreign Trade on the Distribution of Income." *Econ. Tidskr.* 21 (1919).

Hickman, B. G., ed. *International Productivity and Competitiveness.* New York: Oxford University Press, 1992.

Kenen, P.B. "Nature Capital and Trade." *Journal of Political Economy* 73 (1965): 437-60.

———. "Skills, Human Capital, and Comparative Advantage." In *Education, Income, and Human Capital*, edited by W.L. Hanson, 195-229. New York: Columbia University Press, 1970.

Krugman, P. "Pricing to Market When the Exchange Rate Changes." In *Real Financial Linkages Among Open Economies*, edited by S.W. Arndt and J.D. Richardson, Cambrideg, MA: MIT Press, 1987.

Landau, R. "Technology, Capital Formation and U.S. Competitiveness." In *International Productivity and Competitiveness*, edited by Bert G. Hickman, New York: Oxford University Press, 1992.

Leontief, W.W. "Domestic Production and Foreign Trade: The American Capital Position Reexamined." *Econ. Int.* 7 (1954): 3-32.

———. "Factor Proportions in the Structure of American Trade: Further Theoretical and Empirical Analysis." *Review of Economics and Statistics* 38 (November 1956): 386-407.

Linder, S.B. *Trade and Trade Policy for Development.* New York: Frederick A. Prager, 1967.

Monke, E.A. and S.R. Pearson. *The Policy Analysis Matrix for Agricultural Development*, Ithaca: Cornell University Press, 1989.

Ohlin, B. *Interregional and International Trade.* Cambridge, MA: Harvard University Press, 1933.

Piercy, N. *Export Strategy: Markets and Competition* London: University of Wales Institute of Science and Technology, George Allen & Unwin, 1982.

Porter, M. *The Competitive Advantage of Nations.* New York: The Free Press, 1990.

Reca, A., and P.C. Abbott. "Market Access and Agricultural Policy Reform: The Case of European Community Beef Trade." Occasional Paper 31 NC-194, February, 1992.

Sharples, J., and N. Milham. *Longrun Competitiveness of Australian Agriculture.* USDA Economic Research Service, Foreign Agricultural Economics Report No. 243, December 1990.

Solow, R.M. "Technical Change in the Aggregate Production Function." *Review of Economics and Statistics* 39 (1957): 312-20.

———. *Capital Theory and the Rate of Return.* Amsterdam: North Holland, 1973.

Vanek, J. "The Natural Resource Content of Foreign Trade, 1870-1955 and the Relative Abundance of Natural Resources in the United States." *Review of Economics and Statistics* 41 (1959): 146-53.

Vernon, R. "International Investment and International Trade in the Product Cycle." *Quarterly Journal of Economics* 80 (1966): 190-207.

3

A Framework for Assessing National Competitiveness and the Role of Private Strategy and Public Policy

Erna van Duren, Larry Martin, and Randall Westgren

Introduction

Initiatives by government and the private sector to study competitiveness are ubiquitous, yet there is no single conceptual definition that guides these efforts. In fact, one often imputes contrapuntal motives to governments and firms that imply an antagonism between these two groups in searching for competitiveness. This causes competitiveness to be defined differently for firms and nation states, usually on an *ad hoc* basis. Similarly, much of the research on the subject relies on project specific frameworks and models.

Following the Canadian Task Force on Competitiveness in The Agri-Food Industry and its successor Council, we have adopted a definition of competitiveness that is specific and measurable: *Competitiveness is the sustained ability to profitably gain and maintain market share.* This definition has three important and measurable dimensions: profitability; market share; and, through use of the word "sustained," a temporal aspect.

A generalizable framework, based on this definition, is developed for analyzing, measuring and developing strategy for attaining sectoral competitiveness. In doing so we attempt to account for three complicating factors. First, individual firms make the strategic decisions that determine competitiveness of sectors or industries.[1] Hence, there is a need to be able to aggregate firm-specific actions and results to industry level measures. Second, while firms make the decisions, they are affected by public policy. The latter affects strategic choice on a wide range of variables. Hence, a conceptual framework must recognize the interrelationship between private strategy, and public policy. Third, economists are interested in measuring underlying processes based on theoretical concepts. Business people make decisions about relatively detailed activities. The scientific literatures of economics and business reflect the differences in focus on operational detail and underlying processes. The confluence of the "fine-grained" analysis of firm strategy and the prediction and generalizability of economic models are mutually reinforceable. Any useful framework for understanding and measuring competitiveness needs to incorporate both sets of scientific concepts.

A central problem with modelling industry competitiveness is the confounding effects of: (1) the decisions made in the task environment[2] of agricultural industries result from private sector business strategies and public sector policies, and thus are not coordinated; and (2) business strategies and some task environment variables are observable at the individual firm level, while the impact of sectoral policies are observable at the industry-level of aggregation. This leads to what Castrogiovanni calls the over-abstraction problem, as researchers are forced to impute causal relationships of industry-level policies when examining firm-level strategies without being able to establish empirical causality. Even more common is the reverse over-abstraction: researchers impute industry-level effects to business strategies without knowing the effect of industry policies on firm strategy. Any framework for analyzing industry or sectoral competitiveness suffers from this over-abstraction unless it accounts for the strategies of firms.

Following from the foregoing, this chapter:

- crafts the logical arguments for the need to develop a conceptual framework for the measurement, analysis and attainment of industry and sector level competitiveness based on the concepts drawn from both economics and strategic management; and
- develops a generalized conceptual framework for the measurement, analysis and attainment of sector level competitiveness.

Emerging Concepts of Competitiveness and Strategy

The assessment and attainment of competitiveness is a societal and business concern, and one that is the recipient of considerable research and operational effort. At the societal level, competitiveness is concerned with how private firms, quasi-public agencies and governments can co-ordinate their efforts to ensure that living standards are increased, or at least maintained, for as many people as possible. At the level of the firm, competitiveness implies making the necessary strategic and tactical decisions to attain performance goals, such as market share, profitability, growth in net worth, etc. In the extreme, inability to be competitive means business failure. Most of the discussion and analysis on competitiveness deals with these issues in some way. This section reviews major concepts and strategies for creating competitiveness that exist or are emerging from the literature.

Economics Approaches

Many researchers who have addressed competitiveness have been motivated by determining the competitiveness of a given firm, sector or country, and usually at a given point in time. Consequently, most of their definitions and research approaches are static and measurement oriented. In addition, most of their definitions flow from the neo-classical economic theory of trade which postulates that a country will be export competitive in some homogenous product if it has a relative cost advantage in producing and marketing that product.

Comparative advantage is a useful theoretical tool, but it suffers in use from two problems. First, it typically is applied to production costs and not to the totality of marketing, information management and product development costs associated with modern, fragmented, segmented and differentiated output markets. Second, comparative advantage presumes that there is no governmental or "strategic" interest in sustaining domestic production sectors in the face of low cost imports. The political economy of globalized markets will always engender anomalistic sectors protected from the law of comparative advantage.

Cost Oriented Measures. Although the theory of comparative advantage is based on comparative costs, there are no studies of international competitiveness that apply the theory correctly. Several researchers have used differences in costs among nations as indicators of relative competitiveness (West, Arto, Hunt, Tweeten and Pai, Young and Lawson). Unfortunately, most of these studies do not use comparative cost data, but rather conduct a flawed cost comparison based on differences in absolute costs. In some cases this flawed approach is aggravated by comparing the differences in costs of selected inputs. For studies of the food processing industry, raw agricultural products and labor are popular targets for this type of analysis. Several studies use prices to represent cost; either in the domestic market or the landed price in a foreign market. In this approach, exchange rates are often also a critical element in a cost comparison.

The use of resource cost coefficients, the cost of producing a unit of output based on survey and/or engineering studies of input prices and productivity, appears to be the least distorted application of the comparative cost model inherent in the theory of competitive advantage. This approach has been used by Tweeten and Pai, Martin et al, van Duren and Martin, but all of these studies only calculate resource cost coefficients for products in the industry being analyzed. By neglecting other industries, the resulting resource cost coefficients produced in these studies remain closer to the absolute advantage model than to the comparative advantage model. In addition, the inclusion of factors whose prices are a positive function of profitability (e.g., land values) in cost calculations can further distort them.

Productivity Oriented Measures. Underlying the comparative cost variables that drive the comparative advantage model are input prices and the joint productivity of these inputs in producing a unit of output. The output is asserted to be the correct type of output, management practices are assumed to be correct and, consequently, productivity is largely concerned with efficiency, and mostly productive efficiency. Several studies examine productivity, but few do so completely correctly (Hunt, West). Most focus on labor productivity, and only pay lip-service to multifactor productivity.

Economic theory is partly to blame for this state of affairs. Measurement of productivity in real firms, industries and nations cannot conform to the assumptions inherent in the theory. For example, in many cases, the output being produced does not fit with consumers' preferences; during the early 1980s, productivity in U.S. auto plants was largely irrelevant on a per unit basis because the output units had the wrong characteristics. Economic theory tends to assume management is rational, has full information and produces the right product. It often is not so.

The problem is similar, but not identical, to poor pricing efficiency but little has been done to build on that work during the past twenty-five years. Applied researchers, however, commit greater transgressions than economic theory because they focus predominantly on measuring the productivity of individual inputs, and not multifactor productivity. As is the case in cost studies, the productivity of labor is a popular choice. The crucial link to capital is ignored in a surprisingly high proportion of studies, and dealt with inadequately in the studies that do analyze variables such as capacity utilization, economies of scale, etc. (West, Hunt).

Trade Pattern Measures. The theory of comparative advantage indicates that a cost advantage in one product relative to another will lead toward specialization in that product, which is accompanied by increased market share and, in the pure theory, increased exports (or reduced imports). Many studies have used data on trade patterns to draw inferences about competitiveness.

Several export oriented indicators of competitiveness have been widely used. Porter's *Competitive Advantage of Nations* focused on 10 countries that accounted for over 50% of world exports in 1985. Within those countries, the industries chosen for study accounted for a substantial share of exports, were diversified by export destination and had sustained levels of exports (Porter 1990). Competitiveness analysis conducted for the OECD also focuses on exports; specifically, growth of export markets and changes in export market shares as a result of countries' price-competitiveness (Durand and Giorno). Studies on the competitiveness of the agri-food industries conducted on behalf of Agriculture Canada and the U.S. Department of Agriculture have also emphasized trade, more specifically, net export performance (West, Hunt, Vollrath). Generally, industries that have better export performance relative to others in their country and the same industries in other countries are judged to be the most competitive.

Some researchers recognize the role and importance of imports in trade performance and competitiveness (Durand and Giorno, West, Hunt). The import penetration ratio is a useful indicator of the domestic industry's performance in an import defensive industry, while the export orientation ratio is better suited to export oriented industries. The net export orientation ratio recognizes the importance of two-directional trade, and can provide additional insight into an industry's performance, particularly if it is used in conjunction with export orientation and import penetration ratios.

Market Share Measures. Most studies of competitiveness use indicators of trade performance to make inferences about market shares instead of using market shares. One of the interesting exceptions is the U.S. International Trade Commission, which as early as 1985 used the Canadian hog and pork industry's share of the North American market as an indicator of its competitiveness (which it asserted to be the result of subsidization) (Meilke and van Duren). This approach can provide better insight into relative performance than trade ratios, especially if analyzed in conjunction with the change in the size of the total market.

Profit Measures. Although profits are both the result and reason for market share, profitability measures are used in few studies of competitiveness. When they are, profitability is generally represented by gross or net margins, indicators of price-cost behavior and value added measures (USITC, CITT, van Duren and Martin, Martin et al). The numerous studies of firm performance, or profitability, and market share have revealed no generalizable causal relationship, but rather industry specific patterns (Scwalbach, Porter, Chussil) and caution that higher market share does not necessarily guarantee higher profits.

Summary. The research efforts on competitiveness that flow from the neo-classical model are constrained by the static nature of the theory of comparative advantage and the data available to conduct empirical analysis, but they are based on testable hypotheses and provide results accordingly.

Strategic Management Approaches

The strategic management area of the business school's field of inquiry has produced a different approach to competitiveness research. Strategic management begins with the firm as the primary unit of analysis, but explicitly considers organization theories based upon decision-making coalitions within the firm. It also has a basis in contingency theories of organization, which posit that the operating environment of an industry drives firm-level decisions. Much of the evolution of strategic management theory and research is concerned with reconciling the dynamics of decision making with the effects of environmental factors exogenous to the firm.[3] Firm level models, leading to national aggregation, for analyzing competitiveness are represented by three of Michael Porter's books: *Competitive Strategy* (1980), *Competitive Advantage* (1985) and the *Competitive Advantage of Nations* (1990), although numerous other studies continue to address each stage.

Types of Firm Level Strategy and Performance. The strategic management approach to analyzing competitiveness starts with the firm. Nations do not produce, market and trade; individual firms within nations do, and when they do so successfully they are considered to be competitive. For firms to be competitive, they have to develop and pursue effective strategies. Porter's major contribution to the analysis of firm level strategies for competitiveness is the development of three generic strategies for obtaining positional advantage in an industry that can lead to superior performance. Positional advantage is obtained by providing products with associated service bundles for unified or segmented markets that contain specific value for buyers. Above average returns indicate superior performance for each of the generic strategies.

Overall Cost Leadership. This is the approach that many government analysts, most labor organizations and most lay people, continue to associate with competitiveness is the approach that led to success for most "western" firms in the post world war II period. Cost leadership requires an integrated dedication to reducing costs throughout the firm, so that profit margins can increase correspondingly. Efficient scale facilities, vigorous pursuit of cost reductions from the experience curve, strict overhead cost control and a cost-cutting philosophy with core (procurement, operations, marketing) and non-core activities (research and development, finance and personnel and firm infrastructure) are required for success with this approach. In addition to above average returns, decreases in costs and increases in market share in an industry are used as indicators of superior performance with this strategy (Porter 1990, Chussil, Schwalbach, others).

Differentiation. This approach is based on creating a product that is perceived to be unique on an industry wide basis. The best differentiation strategy continuously differentiates on as many facets of the firm's products and processes (goods, services, internal processes) as possible since this makes it more difficult for competitors to copy its formula for success. Although costs are not the primary target, they are not ignored since they continue to contribute to profitability. Since differentiation is frequently based on consumers' perceptions of exclusivity, competitiveness, and hence returns, are not driven by sales volume in this strategy (Porter 1980, Peters 1989).

Niche Strategies. Strategies are based on finding and exploiting a segment of an industry more effectively than firms which operate on an industry wide basis using overall cost leadership or differentiation strategies. The industry segment may be defined on the basis of geographic, product or consumer characteristics. Since a niche strategy is based on limited volumes, high market share does not accompany above average returns as an indicator of superior performance, or competitiveness (Porter 1980, Bradfurd and Ross 1989).

The strongest critique and most useful enrichment of the Porterian approach to firm level strategy for positional advantage comes from Kenichi Ohmae although his views are being amplified by North American strategic management theorists (Day and Wensley, Hamel and Pralahad, Cravens and Shipp). These theorists argue that the industry based, competitor oriented approach to competitive strategy of the Porterian school underrates the importance of serving the customer, and more importantly, of determining what customers may want in the future.

The Value Chain. From among the many models used for empirical studies of strategies, the value chain offers a coherent and simple tool for describing the configuration of firms pursuing any of the generic strategies discussed earlier. The value chain is a firm's system for procuring inputs, producing, marketing, delivering and servicing products, and performing the required support activities. Research and development, personnel, finance and firm infrastructure development comprise a commonly employed framework for translating general notions about strategy into skills, assets, processes and structures that can be used to obtain positional advantage (Porter 1985).

The value chain allows the systematic identification and co-ordination of factors that can be used to pursue a delivered cost, differentiation, niche or superior customer satisfaction strategy. It facilitates examination of the individual factors over which the firm has control, co-ordination of these factors within the firm, and the means to create strategic linkages with other firms. In the short-run, competitiveness results from a value chain that delivers the price and performance attributes that customers want more efficiently than one's competitors. In the long-run, competitiveness results from a value chain that allows the firm to take superior advantage of demands and opportunities and to effectively neutralize threats as they emerge in the business environment (Hamel and Pralahad, Aaker, Porter 1985).

The strategic management literature offers analysis and advice for nearly every function and relationship among functions contained in the value chain. The issue of relationships among firms in an industry is the newest area of research interest; particularly because the Japanese have been very successful with their forms of strategic alliances, their "keiretsu." In order to synthesize this body of research in developing an analytical framework for assessing competitiveness, Table 3.1 provides a brief description of the terms and factors

TABLE 3.1 Factors Commonly Required to Create the Means for Different Strategies for Positional Advantage

Definition and Type of Positional Advantage		Means		
	Assets	Skills	Processes	Structures
Definition and distinguishing characteristics	Tangible and intangible resources that provide services to the firm or are claims to receive income. A possession that is better than your competitor's.	People-based resources that provide services to the firm (knowledge, contacts, etc.). Value is enhanced with use. Something you *do* better than your competitors.	The methods by which assets and skills are combined and decisions regarding the content and combination of business function and strategies are made.	Formal and informal organizational mechanisms to implement, control and adjust the means for obtaining positional advantage.
Overall Cost Leadership	Ability for sustained capital investment.	Process engineering. Applied research capability. Manufacturing and materials management. Basic literacy, numeracy and physical skills among line workers.	Cost control. Products designed for ease in creation. Coordination among various business functions. Linkages with suppliers. Focus on cost/market share/profita-bility in strategic management.	Hierarchical, formal. Manufacturing, finance are relatively more important. Rewards based on "bottom line."

Differentiation	Modern Equipment. Corporate reputation for quality or technological leadership.	Product Engineering. Creativity. Basic research capability and development expertise. Marketing and sales. Basic skills plus problem-solving and conceptual skills among line workers.	Quality control. High quality products. Brands. Coordination among business functions. Linkages with distribution and buyers. Focus on quality and profitability in strategic management.	Problem-oriented work groups. Informal. Marketing research and development are relatively more important. Rewards based on quality and creativity.
Niche	Corporate reputation for quality or technological leadership.	Discovering or creating market niches and opportunities. Appropriate combination of factors listed under overall cost leadership and differentiation.	Focus on maintaining niche, sometimes growth, in strategic management. Appropriate combination of factors listed under overall cost leadership and differentiation.	Appropriate combination of factors listed under overall cost leadership and differentiation.
Superior Customer Satisfaction	Corporate reputation for quality in products, indirectly associated services.	Discovering or creating products and indirect services not yet recognized as good value by customers. Appropriate combination of factors listed under overall cost leadership and differentiation.	Ongoing market research. Focus on the customer in strategic management. Appropriate combination of factors listed under overall cost leadership and differentiation.	Problem/opportunity-oriented work groups. Appropriate combination of factors listed under overall cost leadership and differentiation.

that are relevant to determining how to succeed with the strategies for positional advantage.

National Competitive Advantage. National competitiveness occurs when a sufficient number of firms create the means for sustainable positional advantage, and generate enough profit to finance the private and public sector's role in achieving their responsibilities. The major studies of competitiveness conducted to date focus on the private sector component of this definition, although research on the public policy aspects of the challenge of creating national competitiveness is becoming more popular.

Porter's *Competitive Advantage of Nations* is representative of the research on national competitiveness, and with a few exceptions, this approach remains the most comprehensive. It is an adept extension of the comparative advantage approach to defining, measuring and explaining competitiveness. It posits that competitive advantage can be created, and that certain conditions, which are embodied in his "national diamond model," lead to its creation. Four sets of variables: factor conditions, demand conditions, related and supporting industries, and firm strategy, structure and rivalry, which are influenced by two additional variables, chance and government, form the basis for Porter's research. Based on in-depth case studies of relatively sophisticated, internationally successful or improving industries and industry segments in 10 countries[4] that accounted for over 50 percent of world trade in 1985, Porter's study concludes that the characteristics summarized in Table 3.2 create internationally competitive industries, and thus national competitiveness.

Porter's research, along with the research and numerous commentaries of strategic management researchers (Thain,Scott), places a very high value on creating the means for competitiveness within a nation. It is one of the central themes in Reich's *Work of Nations*, which examines national competitiveness issue from a public policy angle. Critics of Porter's diamond model of national advantage question a number of his assertions. Rugman's criticism is the most pointed and best supported, but it applies mostly to the small country case. He argues that the national diamond is not the relevant issue, but rather that the conditions within the relevant trading block should be considered. In Canada's case the relevant diamond is North American. Consequently, Porter's analysis of foreign ownership is flawed. Rugman argues that it is 30 years out of date and does not draw on the abundance of work by Canadian scholars refuting the negative impact of foreign owned multinationals on the Canadian economy (Rugman, 1991 and 1992).

Other criticisms of Porter's work deal with his assertion that strong horizontal competition fosters greater innovation and efficiency. Many argue that such an arrangement is incompatible with the scale and scope of organizational means required to compete in the world markets for many industries (Spring, Rugman).

Economists generally criticize Porter's research because the results are not

TABLE 3.2 Creating National Competitiveness

Sets of Variables	Characteristics that Lead to International Competitiveness
Factor Conditions	Effective and efficient deployment of factors. Basic factors, which are passively inherited or require relatively modest or unsophisticated private and/or social investment and often can be acquired on world commodity markets, are becoming less important. Advanced factors, which require focused and extensive social and private investments and are less internationally mobile, are becoming more important. Generally, accessible factors are less important than specialized factors that cannot be easily used by other industries. Factor base that is advancing and specializing on a continuous basis is best bet for competitive advantage.
Demand Conditions	Market segment has relatively greater strategic importance in the home market than in world market. Domestic customers are more sophisticated and demanding than their international counterparts. Large home demand is more important in industries and segments in which economies of scale and learning exist. Numerous independent buyers increase the pool of demand information available. Rapid growth rate and early market saturation induce firms to innovate and upgrade more rapidly and intensively. Buyers in the country have a positive impact on international demand.
Related and Supporting Industries	Presence of internationally competitive supplier industries: early, efficient access to most innovative and cost effective inputs; ability to coordinate in product and technology development and create a process of continuous innovation and upgrading. This part of the diamond depends critically on the others.
Firm Strategy, Structure, and Rivalry	Varies considerably by nation due to culture and inertia of existing economic structures. Global attitudes and experience (language, operating, strategic). Flexibility (business norms and practices). Sustained commitment. Strong horizontal domestic rivalry. New business can be established with relative ease.
Government	Policy, programs, and instruments that amplify the characteristics with positive impacts are conducive to sustainable national competitive advantage.
Chance	Events of the following types with positive impacts on the industry: pure invention and technological discontinuities; supply and demand side shocks; and political events beyond the home government's control.

Source: Porter, M. The Competitive Advantage of Nations. New York: The Free Press, 1990.

based on testable hypotheses. This is a weak criticism because Porter's approach recognizes the tradeoff between rigor and relevance. Strategic management research relies more heavily than economics research on in-depth field studies (Schwenk), both to develop the theory of behavior internal to the firm and to test those theories. Economic models generally lack the richness and depth of understanding of complex relationships that are attained through the strategic management researchers' use of case studies. Although case studies are dismissed as anecdotes by some researchers, the results or "divine revelations," they produce information that extends firm-level theories of the formation of private strategy. Combining case studies and economic analyses would begin to reconcile the over-abstraction problems associated with economics research, which relies on highly aggregated data, and the conceptual ambiguity or non-generalizability associated with some strategic management research (Castrogiovanni, Schwenk).

A Conceptual Framework

The development of a framework stems directly from the definition of competitiveness that was presented in the introduction: *the sustained ability to profitably gain or maintain market share.* This definition is broadly consistent with what firms try to achieve, that is high levels of, or growth in, profits or sales. It is also consistent with what governments try to achieve with commercial or economic policy for their sectors.[5]

A Conceptual Model of Competitiveness

Given the definition of competitiveness developed in this paper, business strategy can be broadly defined as taking advantage of those factors over which one has control in order to achieve competitiveness. Firms as well as governments have control over certain factors within the business environment of a given sector. One can construct a model of competitive advantage that includes the causal relationships between the factors over which firms and governments have control and the achievement of competitiveness. Such an approach incorporates some of the arguments of neoclassical economics, industrial organization and strategic management discussed above. Figure 3.1 reflects that the attainment of competitiveness at the national level is a function of factors that are under varying degrees of control and influence of private firms and governments. In general, private firms develop strategy within the environment established by governments, although this environment is the product of pressures from diverse stakeholders, including business. Figure 3.1 provides a useful way of conceptualizing which factors are, can or should be controlled by private firms and government in the quest for national competitiveness. Both firms and government can create the means and acquire the positional advantages required for competitiveness. Table 3.2 contains a selection of our "Porterian conclusions," or rather, strong hypotheses, about which factors are best controlled by firms and which are best controlled by government in order to create competitiveness. The boxes for factors controlled by firms and by government in Figure 3.1 present these factors in summary form.

Figure 3.1 indicates that within the environment established by governments, individual firms can control their strategies; the products they produce; the technology they adopt and, therefore, their internal research and develop-

Competitiveness is indicated by:

Profits	Market Share

Competitiveness is affected by factors that are:

Controlled by the Firm	Controlled by Government	Quasi-Controllable	Uncontrollable
Strategy *Products* *Technology* *Training* *Internal R&D* *Costs Linkages* *Strategic Alliances*	*Business* *Environment:* *taxes* *interest rates* *exchange rate* *R&D Policy* *Regulation* *Education &* *Training* *Strategic Alliances*	*Input Prices* *Demand Conditions* *International Trade* *Environment*	*Natural Environment*

FIGURE 3.1 Indicators of Competitiveness and Explanatory Variables.

ment policies; their cost structures; their processes for human resource development; and relationships with customers or suppliers. These relationships, or linkages, are often the basis for total quality management, joint ventures, and flexible vertical relationships that do not rely upon ownership or formal contractual relationships.

Governments have some degree of control over the general business environment, through fiscal and monetary policies, research and development policy, market structure (through competition policy), education and training policy, and to some extent, the form of linkages among firms or between levels of the market[6]. The latter is particularly relevant in the agri-food industries due to the structure of agricultural policy and marketing boards. When policy promotes antagonism between supplier industries (farming) and customer industries (primary and secondary processors), effective linkages between individual suppliers and buyers cannot be forged. Governments also affect the cost structures, product mixes, marketing efforts and the like through the regulatory environment they develop.

Quasi-controllable factors are those that are beyond the limited control of managers or government. We include input prices in this category because, to the extent that inputs are traded internationally, no company or government has much control over them. Similarly, demand conditions are included because firms have a limited ability to affect them. For example, demand for red meats continues to decline, while demand for white meats continues to increase. Firms, trade associations, and governments have limited ability to encourage demand through product-specific and generic advertising programs. Another class of factors that cannot be completely controlled, that is particularly important in resource based industries, is the impact of the interaction of natural environmental factors such as soil and water conditions with production technologies on the ability of the resource to sustain production at current and improved levels. These quasi-controllable conditions affect firm strategies, conduct, and ultimately performance (i.e., competitiveness at firm, industry, or sector level).

Finally, there are uncontrollable factors. They include those envisioned in neoclassical economics, that is, the endowments of natural resources, geography and climate.

Application of the Model

To complete the development of a model of causal relationships for competitiveness, it is necessary to organize the concepts from Figure 3.1 into systematically measurable variables. These are indicated in Figure 3.2. The top part of Figure 3.2 contains alternative variables for profits and market shares, the variables that represent performance or competitiveness. For the purposes of measuring competitiveness at the level of an industry, such as the SIC 4-digit

Competitiveness is measured from the perspective of and with:

| Profits | | Market Share |

Value Added per:

sales
worker
plant
wage bill
output
time

Trade Ratios:

export orientation
import penetration
net export orientation
trade coverage

Canada's Share of Sales
in North America

Using the following drivers:

Productivity	Technology	Products	Inputs & Costs	Concentration	Demand Conditions	Strategic Alliances

FIGURE 3.2 Variables Used as Competitiveness Indicators and Competitiveness Drivers.

aggregate, value-added is proposed as a proxy for profitability. It is extremely difficult to obtain or use direct measures of profit to represent the industry. This is due to differences in accounting procedures, differences in the sizes of firms (so that the weighting of individual firms within an industry average is extremely difficult), the fact that many companies are not publicly traded and, therefore, do not provide financial statements, and, finally, because multidivisional public companies normally provide only consolidated financial statements.

With all these problems, it is more useful to obtain indirect indicators of profits (Davis, et al.). Revenues are sometimes used as a proxy. While sales give an idea of total activity, they are a poor representation of profits. Value added is better because it indicates the industry's surplus over raw material costs. Value added is also important to governments, which design programs to increase value added on the ground that they perceive it to be an indicator of increased economic activity (especially employment) in domestic industries (Reich).

One must be careful to understand the nature of "value added" when it appears in empirical work. Cook and Bredahl chastise American analysts for assuming that value added is tantamount to increased processing. They agree with the usage of the term in this paper, wherein value added may be increased by decreasing costs or by differentiating products and processes at any and all stages of the market channel. Therefore, it provides at least an initial bridge between the conflicting concepts of the economics and business strategy literatures that were discussed above. In representing profits, a basic approach for an industry is to measure the change or growth rates over time for the domestic industry compared to those in competing countries in order to obtain a view of relative trends.

Value added can also be expressed as a ratio of sales, the number of employees, the cost of labor, or the number of establishments and to show growth in the variables. With multiple ratio measures of value added, the intention is to draw an inference about the overall state of an industry by isolating trends in value added relative to several numeraires associated with fixed and variable inputs.

Similarly, there are alternative variables one could construct for market share. Most obvious is the industry's share of its domestic market or of some international market of relevance (e.g. the North American market). Cook and Bredahl point out that this definition of competitiveness can be usefully applied in measurement of market share in aggregate sectors (e.g. protein market) or in particular time periods, as well as geographical markets. Other measurement variables that have been used in the past include the *export orientation ratio* and the *import penetration ratio*. The export orientation ratio expresses the domestic industry's exports as a percentage of its total production. The import penetration ratio expresses imports from foreign countries as a percentage of domestic consumption. These measures are usually national in scope, but inferences may be drawn for the regional industries' "international" competitiveness. Over time, these measures give a good indication of relative growth.

A variant on the export orientation/import penetration concept is the *net export orientation ratio.* It is defined as the difference between exports and imports expressed as a ratio of the average of domestic consumption and domestic production. It provides a good view of the relative importance of imports and exports and, over time, how the importance changes especially when complemented with the *trade coverage ratio.*

If the foregoing variables indicate the competitiveness of an industry, then what are the factors that contribute to the competitive state? Neoclassical economics, industrial organization economics, and strategic management provide conceptual answers to the question. The causal variables are indicated in the boxes at the bottom of Figure 3.2. They are referred to through the remainder of the report as *drivers* of competitive advantage.

In most cases, these variables are jointly reflective of the four factors (controlled by firms, by governments, quasi-controllable, uncontrollable) listed in Figure 3.1. For example, an industry's cost structure is clearly affected by decisions made by individual firms on the matters listed in Figure 3.1. However, it is also affected by macroeconomic policy, international trade policy and the regulatory environment, all of which are controlled by government. Costs may also be affected by quasi-uncontrollable factors such as international markets and uncontrollable factors such as natural resource endowments and climate. These factors jointly affect the manner in which cost structure drives an industry's competitive state.

Similar arguments can be made about the joint responsibility of business and government for all seven drivers at the bottom of Figure 3.2. The question is how can they be measured? For five of the seven drivers, measures may be constructed from secondary data. In addition, it is possible for all seven factors to obtain at least general measure through what we call strategic assessment: primary data from strategic managers.

The first driver is productivity. Productivity can be measured by labor productivity, data which are readily available, or, as indicated earlier, by data on multi-factor productivity, which are more difficult to obtain because so few analysts calculate them.

The second driver is the use of technology. A direct measure of technology is expenditure on research and development, both the level of expenditure and the change in real expenditure. However, given that much technology (assuming that technology is embodied in capital) in agriculture and food industries is obtained by importation and not by domestic research and development, this measure misses much of the technological change that occurs. A second but less direct measure of technology is, therefore, to express the capital-labor ratio for the industry over time. If embodied technology is growing, then one would expect the capital-labor ratio to increase.

The third driver is the number, quality and appropriateness of products. There is no good measure from secondary sources of the ability of an industry to develop products that enhance its competitiveness. From the marketing perspective, the ability of existing or new products to compete is determined by a number of factors. These include the convenience of the product or its packaging, the effect of the product or its manufacturing process on the environment, quality, nutritional value, and safety to the consumer. There is no source of secondary data that represent these drivers. Hence, the only approach is to obtain primary data through strategic assessment.

The fourth driver is inputs and costs. The primary concern is with the relative prices of major inputs and ingredients. The first variable is simply a comparison of prices of major inputs among competing countries. The second is the *resource cost coefficient*. Resource cost coefficients represent the unit cost of production for representative products in different countries. For example, one might assume a particular set of technical coefficients to produce bread and then calculate, for various countries, the costs of production using this process. This shows not only the differences in prices of inputs, but also their relative impacts. For example, if a particular process requires two inputs and the price of one in Canada is higher than in the United States, but vice-versa for the second, the resource cost coefficient shows whether the higher price is offset by the lower. It also focuses, where input price differentials are caused by government policy, on policy changes that need consideration.

The fifth driver is intended to bring into the argument some of the concerns of industrial organization. In addition, it represents issues regarding competition policy. Our suggested variable is the concentration ratio of the top four firms (CR_4). There are many problems with the use of CR_4 as an indicator of monopolistic practices, but the ratio is a basic indicator of relative scale.

The penultimate driver is demand conditions. As has been indicated previously, this can be a very important factor in determining the strategy of firms and in determining their competitiveness. Longitudinal measurement of physical quantities of consumption and cross elasticities computed for substitute products are indicators of the demand driver.

The final driver is alliances. This represents the interface between firms and their suppliers and customers. Alliances can relate to anything from the price formation process between growers and processors, to co-packing arrangements, to the ways retail shelf space is obtained, to the types of customer service a company provides.

There are no secondary sources of quantitative information on linkages, at least at the level of the industry. Those arrangements which are based upon strategic interactions between supply firms and processors and that are not based upon ownership (vertical integration) or contracting are extremely difficult to identify and quantify in practice. Among these relationships would be total quality management agreements and just-in-time inventory systems, which are important to the cost and product quality dimensions of firm performance, but difficult to isolate in empirical analysis.

Conclusions, Applications, and Potential Extensions

Several points can be made regarding the framework developed here based on experience with it to date. First, the actual measures that have resulted and been reported elsewhere (Martin, et al.) appear to be valid. They have been presented in detail to senior managers in several of the industries and there is agreement that they present a reliable picture of their competitive state. In particular, the cross relationship among the several value added ratios seems to have appeal. If they all point to the same conclusion, the reaction is that they are rather overwhelming in their message. If they are mixed, that is, if they indicate that the Canadian industry is superior on one and the U.S. industry is superior on another, the reaction is generally that they accurately reflect inherent differences. The authors conclude that the framework is a valid one for assessing a processing industry's competitive state. Our sole reservation lies in applying the framework to Canada's supply managed industries. The market share measures are relatively meaningless since the industries are protected by quantitative import restrictions. Moreover, there is a suspicion that value added reflects market power resulting from the protected environment. If so, Canadian industries may be assessed more positively than they deserve. The arguments are not simple and, in our view, the jury is still out on this issue.

Second, the framework provides an advantage over earlier models of competitiveness advanced by the business strategy literature. It is underpinned by accepted economic principles. It adds the richness brought by the reality of firm-level strategy. Most importantly, it goes beyond the models of Porter and others by responding to the criticism that Porter's model does not deal explicitly with the role of government. The current framework explicitly and separately recognizes that both government and private industry have a role to play in achieving competitiveness. It also recognizes that the two jointly affect what we have termed the drivers of competitiveness through the interaction of the public policy arena.

Third, the framework is proving to be quite useful for diagnosing problems and establishing priorities for reforming policies affecting the competitiveness of Canada's agri-food industries. It has been used by the Agri-Food Competitiveness Council in developing several recommendations that have been adopted as federal and/or provincial policy on taxation and regulation. Underlying this is the fact that it has contributed to developing the Council's priorities as reflected in its work plan. The latter has recently been jointly endorsed by the federal and provincial Ministers of agriculture. An interesting test of the development and ordering of priorities that it provides is to compare the priorities established by the Council using the framework to those identified by a broader cross-section of the sector through the consultative process associated with the federal "prosperity initiative." They are almost identical. Several other industry reports are now being undertaken using the framework (van Duren), and a number of industry groups are using the framework as the basis for providing the information needed to undertake long term strategic planning.

Fourth, there are underdeveloped relationships between the framework and economics that could use attention. The foregoing point indicates that the framework is useful in establishing priorities in changing policy. But this is not done in the formal sense of calculating benefit-cost ratios, but rather through the experience and judgement of senior people responsible for the development and application of strategy. It is difficult to see how, in the limited time horizon and limited information environment that characterizes today's political economy, formal analysis of a timely nature can be done. However, it is an area that deserves attention. The same can be said for the application of welfare concepts. There are several such extensions that could make the framework more theoretically sound.

Notes

1. We distinguish between sectors and industries as follows. An industry is a three or four digit SIC category (the beef processing or red meat processing industry). A sector is a grouping of closely related SIC categories, their raw product suppliers and (where relevant) their specialized distributors (the agri-food sector).

2. The task environment is defined as the competitors, customers, suppliers, and regulatory bodies that directly affect organizational goal attainment (Bourgeois, 1980). If the goal is competitiveness, then a complete and explicit analysis of the task environment is engendered in the firm-level decisions of rivals, buyers, related and supporting industries, suppliers of factors, and government policies specific to the industry.

3. See Astley and van de Ven for a discussion on the perspectives on management.

4. The 10 countries are United States, Japan, West Germany, United Kingdom, Italy, Sweden, Switzerland, Denmark, Korea and Singapore.

5. The intention, with the use of "profitably" is to imply that market share is gained or maintained by use of legitimate business practices. Manipulation of exchange rates or the use of government subsidies is not what is intended. More importantly, it implies the sustainability of competitiveness; over the long run, the unprofitable industry fails.

6. Levels refer to producers, primary processors, secondary processors, distributors and retailers.

References

Aaker, D.A. "Managing Assets and Skills: The Key to a Sustainable Competitive Advantage." *California Management Review* 31 (Winter 1989): 91-107.

Arto, E.W. "Relative Total Costs — An Approach to Competitiveness Measurement of Industries." *Management International Review*. 27 (1987): 47-58.

Astley, W.G., and A.H. van den Ven. "Central Perspectives and Debates in Organization Theory." *Administrative Quarterly*. 18 (1983): 245-73.

Badaracco, J. Jr. "Alliances Speed Knowledge Transfer." *Planning Review* 19(March/April 1991): 11-16.

Becker, G.S. *Human Capital*, New York: Columbia University Press, 1964.

Bradfurd, R.M., and D.R. Ross. "Can Small Firms Find and Defend Strategic Niches?" *The Review of Economics and Statistics* 71(1989): 258-62.

Bressler, R.G. *Markets, Prices, and Interregional Trade*. New York: John Wiley and Sons, 1970.

Buchanan J.M. "An Economic Theory of Clubs." *Economica* 32(1965): 1-14.

Castrogiovanni, G.J. "Environmental Munificence: A Theoretical Assessment." *Academy of Management Review* 16 (1991): 542-65.

Chussil, M. "Does Market Share Really Matter?" *Planning Review* 19(September/October 1991): 31-7.

Cook, M., and M.E. Bredahl. "Agri-business Competitiveness in the 1990s: Discussion." *American Journal of Agricultural Economics* 73 (December 1991): 1472-73.

Cravens, D.W., and S.H. Shipp. "Market-Driven Strategies for Competitive Advantage." *Business Horizons* 39 (January/February 1991): 53-61.

Davis, E. *et al.* "Who are World's Most Successful Companies?" *Business Strategy Review* 2 (Summer 1991): 1-34.

Day, G.S., and R. Wensley. "Assessing Advantage: Framework for Diagnosing Competitive Superiority," *Journal of Marketing* 52 (April 1988): 1-20.

Douglas, S. P., and Y. Wind. "The Myth of Globalization." *Columbia Journal of World Business*. 22 (Winter 1987): 19-29.

Douglas, S. P., and C.S. Craig. "Evolution of Global Marketing Strategy: Scale, Scope and Synergy." *Columbia Journal of World Business* 24 (Fall 1989): 47-59.

Drucker, P.F. *The New Realities*. Harper and Row: New York. 1989.

———. *Management*. New York: Harper and Row, 1973.

Durand, M. and C. Giorno. "Indicators of International Competitiveness: Conceptual Aspects and Evaluation." *OECD Economic Studies*. 12 (Autumn, 1989): 147-82.

Hamel, G., and C.K. Prahalad. "Corporate Imagination and Expeditionary Marketing." *Harvard Business Review* (July/August1991): 81-92.

Hunt L. "The International Competitiveness of Canadian Fruit and Vegetable Processing Industries." *Food Market Commentary* 8.

Just, R.E., D.L. Hueth,. and A. Schmitz. *Applied Welfare Economics and Public Policy* Englewood Cliffs, N.J.:Prentice-Hall, 1982.

Lambkin, M., and G.S. Day. "Evolutionary Processes in Competitive Markets: Beyond the Product Life." *Journal of Marketing* (July 1989): 4-20.

Lancaster, K.J. "Change and Innovation in the Technology of Consumption" *American Economic Review* 56 (1966): 14-23.

————. "A New Approach to Consumer Theory," *Journal of Political Economics* 74 (1966): 132-57.

Levitt, T. "The Globalization of World Markets." *Harvard Business Review* 61 (May/June, 1983): 92-102.

Martin, L., R. Westgren, and E. van Duren. " Agribusiness Competitiveness Across National Boundaries," *American Journal of Agricultural Economics* 73 (December, 1991): 1456-71.

Martin, L., E. van Duren, R. Westgren, and M. Le Maguer. "Competitiveness of Ontario's Agrifood Sector." Prepared for the Government of Ontario, May 1991.

————. "Competitiveness of Food Processing in Canada." Prepared for Industry, Science and Technology Canada, Food Policy Task Force, May 1991.

Moss-Kanter, R. "Transcending Business Boundaries: 12,000 World Managers View Change," *Harvard Business Review* (May/June, 1991): 151-64.

Ohmae, K. "Getting Back to Strategy." *Harvard Business Review* 66 (November/December, 1988): 149-56.

Peters, T., "New Products, New Markets, New Competition, New Thinking." *The Economist* 36 (March 4, 1989): 19-22.

————. "Prometheus Barely Unbound" *Academy of Management Executive* 4 (1990): 70-84.

Popcorn, F. "What If She is Right?" *Report on Business Magazine* (October, 1991): 66-77.

————. *The Popcorn Report*, New York: Doubleday Currency, 1991.

Porter, M. "How Competitive Forces Shape Strategy." *Harvard Business Review* 57 (March/April, 1979): 137-45.

————. *Competitive Strategy: Techniques for Analyzing Industries and Competitors* New York: The Free Press, 1980.

————. *Competitive Advantage: Creating and Sustaining Superior Performance* New York: The Free Press, 1985

————. *The Competitive Advantage of Nations* New York: The Free Press, 1990a.

————. "The Competitive Advantage of Nations." *Harvard Business Review* 68 (March/April, 1990): 73-93.

Porter, M., J. Armstrong, D. Thain, and P. Ballinger. "Canada at the Cross-Roads." *Business Quarterly* 57 (Winter 1992): 6-15.

Porter, M., and V.E. Millar. "How Information Gives You Competitive Advantage." *Harvard Business Review* 64 (July/August, 1986): 149-58.

Pralahad, C.K., and G. Hamel. "The Core Competence of the Corporation." *Harvard Business Review* 67(May/June, 1989): 79-91.

Reich. R. *The Work of Nations*. New York: Alfred A. Knopf, 1990.

Rugman, A., and A. Verbeke. "Trade Barriers and Corporate Strategy in International Companies — The Canadian Experience." *Long Range Planning* 24 (1991): 66-72.

Rugman, A.M. "Diamond in the Rough." *Business Quarterly* 57 (Winter, 1991): 61-4.

————. "Porter Takes the Wrong Turn." *Business Quarterly* 56 (Winter, 1992): 59-64.

Saatchi and Saatchi. *The Consumer of the 1990's.* 1990.

Schwalbach, J. "Profitability and Market Share: A Reflection on the Functional Relationship." *Strategic Management Journal* 12 (1991): 299-306.

Scott, B.R. "Competitiveness: Self Help for a Worsening Problem." *Harvard Business Review* (July/Aug, 1989): 115-21.

Spring, D. "An International Marketer's View of Porter's New Zealand Study." *Business Quarterly* Winter 1992): 59-64.

Toffler, A. *Power Shift: Knowledge, Wealth and Violence at the End of the 21st Century.* New York: Bantam Books, 1990.

Tweeten, L., and D. Pai. "Public Policy and the Competitive Position of U.S. Agriculture in World Markets." *Organization and Performance of World Food Systems* Occasional Paper Series, NC-194, March 1990.

U.S. International Trade Commission. *Live Swine and Pork from Canada.* Pub. 1733, July 1985.

Van Duren, E. "Government Policy and the Competitiveness of Ontario Vegetable Processors and Growers." Submission to the Canadian International Trade Tribunal's Inquiry into the Competitiveness of Canada's Horticulture Industry, 1991.

Van Duren, E., and L. Martin. "Assessing the Impact of the Canada-U.S. Trade Agreement on Food Processing in Canada: An Analytical Framework and Results for Poultry, Dairy and Tomatoes." *Agribusiness: An International Journal* 8 (January 1992): 1-22.

Van Duren, E. L. Martin, and R. Westgren. "Assessing the Competitiveness of Canada's Agrifood Industry." *Canadian Journal of Agricultural Economics* 39 (December 1991): 727-38.

Vollrath, T.L. *Competitiveness and Protection in World Agriculture.* USDA Economic Research Service, Agriculture Information Bulletin, No. 567, July 1989.

West, D. "Productivity and the International Competitiveness of the Canadian Food and Beverage Processing Sector." *Food Market Commentary* 9 1(987).

Willig, R.D. "Consumer Surplus Without Apology." *American Economic Review* 66 (September 1976): 589-97.

Young, H., and A. Lawson. "Exchange Rates and the Competitive Price Position of U.S. Exports and Imports." *Business Economics* (April 1988): 13-9.

4

Innovation, Entrepreneurship, and Non-price Factors: Implications for Competitiveness in International Trade

Thomas R. Pierson and John W. Allen

Introduction

What can food companies do to enhance success in international business? Surely food executives need to be "innovative" in grappling with the challenges of extending businesses beyond known geographical territories and conventional methods. Perhaps the most important and difficult challenge is that of responding corporately to customers and consumers possessing different cultural, demographic, and lifestyle characteristics and the different ways by which they purchase and consume foods. Responses by companies to expand internationally may involve one or more approaches: the shipment of products to foreign markets, the establishment of foreign production facilities, the use of licensing agreements, entry into joint ventures/partnerships, or perhaps the acquisition of other firms. Clearly, "entrepreneurship" is a managerial quality important to achieving success in these approaches. While acknowledging the importance of "price," this chapter emphasizes the many factors beyond price that are essential to successfully market products in a global setting. Despite the importance of price in international trade, especially in a setting of faltering economic growth and overproduction of many foods, price is but one among several factors that determine customer and consumer acceptance of product offerings. It would be presumptuous to assert that there exists a definitive formula for success in the process of developing international businesses. Nor is it reasonable to expect that international business can ever be free of difficulties or risks even when emphasis is given to non-price factors. Yet it is clear that many companies of significantly different sizes and characteristics are achieving success by appropriately balancing price and non-price factors.

International markets for many companies are now major sources of profit and, for a growing number of firms, a strategic priority for future growth and profitability. This is true despite the current worldwide recessionary trend that affects to a greater or lesser extent the economies of the most developed nations in the world. These weakened economies have caused setbacks to the plans and progress of many companies as they strive to expand abroad. Other global challenges are causing executives to pursue international business plans with caution: the tendency of global trading blocs, which may diminish the hoped-for trend toward worldwide free trade; the problems encountered in the unifying developments of the European Community; the difficulties associated with the transformation to market economies of the Commonwealth of Independent States and Eastern Europe. China, India, and the Asian Pacific Rim nations, with their enormous populations, each possess complex economic, social, and political situations resulting in trade uncertainties, despite their overall compelling potentials for trade.

Whatever the challenges and difficulties, there is a consensus among perceptive food executives that not to become engaged in international business is not a viable option. Moreover, it is not only large companies that recognize the need to be engaged in the global marketplace. Small firms, as well, must strategize for their respective places in international trade. Consider this viewpoint expressed by the editor of *Food Business*: "...any company that doesn't look to foreign markets for future growth is asking for trouble in a world that is quickly becoming globalized. To any small to mid-sized company, I say: You are not safe in America" (Messenger, 1991).

Advantages accruing to foreign initiatives often contribute beyond direct sales and profits. Successful business abroad can enhance respect and credibility among traditional, domestic customers and other domestic firms that are important for success. Beyond gaining an enhanced business image is the advantage of being informed of emerging product, industry, social, and economic developments that often have their origins in other areas of the world.

The Meaning of Marketing

Fundamental to the success of any food company operating in this increasingly global setting is to fully grasp the contemporary meaning of marketing. Traditionally, the term marketing was narrowly regarded as the act of shipping food and agricultural commodities to the marketplace. Today food marketing has taken on a much broader meaning for many food and agricultural organizations. The term now encompasses the process of adding value to farm and food products, providing ultimate consumers and customers throughout the food system with products and services that meet desired wants and needs.

Contemporary marketing has become essential for profitable sales across the global food and agriculture system. Traditionalism is on the decline as more effective marketing comes of age in the sense that consumers, retailers, wholesalers, and processors as well as shippers, assemblers, producers and growers are changing their attitudes and behaviors with respect to food and agribusiness. More food companies are developing marketing strategies linked to the following basic components of the marketing process:

- knowledge and understanding of ultimate consumers' and intermediary customers' wants, needs, and perceptions;
- development and positioning of products and services to match customers' wants, needs, and perceptions; and
- communication of positive product and service attributes to targeted customers.

Although this list of the basic components of marketing understates the complexity and needed creativity of the marketing process, it indeed contains the central core of contemporary marketing processes that are vital for success in today's international food and agricultural industries.

On an aggregate commodity basis, in the most highly developed nations, food consumption rises slowly, about equaling population increases. However, food products of higher quality that meet more demanding specifications for size, color, condition, maturity, tenderness, consistency, lower bacteria counts, less saturated fat, etc., have witnessed dramatic increases in sales. Conversely, undifferentiated products and commodities, especially those of ordinary quality sold for traditional markets and uses, chronically overwhelm demand, frequently forcing prices below profitable levels. "Market driven" food and agricultural organizations avoid the pitfalls of the commodity-oriented approach by responding to emerging consumer and customer trends and preferences, which in turn create potentially profitable situations throughout food production and distribution (Pierson and Allen, 1992)

Although many food and agricultural firms are market-driven, all too many remain production-, sales- and commodity-oriented; the latter organizations focus primarily on producing and selling commodities, rather than marketing efforts, to determine product attributes and services that customers and ultimate consumers want and are willing to pay for. Though the basic customer-oriented marketing process described above is simple in concept, its successful implementation requires tremendous commitment, creativity, and high levels of skill.

Frequently food and agricultural firms confront traditional markets with burdensome supplies of products and services, most often of ordinary quality for conventional uses. As a result, these undifferentiated products almost always yield low prices and profits except in times of scarcity. Thus these firms experience variable, but typically sub-par, profitability. Fortunately many firms throughout the food system are beginning to realize that greater profit opportunities are often generated by developing and marketing products that are in step with changing customers and that their customers and ultimate consumers are willing to pay premiums for products possessing benefits superior to those of undifferentiated commodities.

Thus, the application of the marketing process is perhaps the most important positive development taking place in the global food system. Industry leaders are turning away from a traditional commodity/production orientation in favor of marketing based upon customers and consumers' wants and needs. These differing philosophies are characterized by the following business approaches:

Commodity/Production Orientation	*Marketing Orientation*
Production and sales of existing products are maximized, often over-looking customer needs and related profit opportunities. Customers are usually viewed as being alike, and the marketplace is perceived as static. Appeals to customers are based primarily upon prices and quantities of traditional commodities.	Customer satisfaction is maximized with products and services based upon wants and needs. Customers are viewed as different from one another with differing and identifiable wants and needs.

Admittedly, there are enormous challenges confronting contemporary marketers in their efforts to implement the marketing orientation, as the difficulty of marketing tasks escalates. However, to remain production and commodity-oriented is not a viable long-range strategy. Commodity sellers must be aware that the growing wave of market-oriented products is creating increasingly formidable competition in both domestic and global settings. This challenges management to acquire the skills, resources, and energy in order to survive and prosper in a marketing oriented framework. This reality is not only the challenge; it is the singular opportunity.

Consumers, Customers, and Purchasing Behavior

Fundamental to the success of food companies in marketing is the analysis of consumers' wants and needs. This provides the basis for creating products, services, and distribution systems that truly meet the needs of targeted consumer segments. Experienced marketers stress the importance of recognizing the distinguishing characteristics of consumer groups, especially in a global context. There are few, if any, food products that can be successfully sold across nations without at least some modification. Even such a global "super brand" as Coca Cola must be marketed in each country with at least some product and / or distribution adjustments, even though the soft drink formula remains unchanged.

A useful framework for food marketers to apply in their analyses of consumers in the national or global marketplace is a set of purchase criteria that consumers use in determining whether a given food product meets their wants and needs. Consumers judge the benefits of all food products and services with a dynamic set of criteria, which changes over time and which differs among individuals. The following words and phrases are used in today's marketplace by consumers to describe and assess food products and services:

Appetizing Appearance	Consistency	Convenience
Environmentally Compatible	Socially Responsible	Excitement
Freshness	Nutrition	Quality
Safety	Taste	Trust
Information	Distinctiveness	Value
Variety	Price	

In today's highly competitive arena, if products and services are to succeed over time, they will likely meet several important consumer wants and needs. It is instructive to consider the following consumer purchase decision criterion that applies to food products:

$$\frac{\text{Consumer}}{\text{Value}} = \frac{\text{Benefits}}{\text{Price}} = \frac{\begin{array}{lll}\text{Appearance} & \text{Consistency} & \text{Convenience}\\ \text{Environment} & \text{Social} & \text{Excitement}\\ \text{Freshness} & \text{Nutrition} & \text{Quality}\\ \text{Safety} & \text{Taste} & \text{Trust}\\ \text{Information} & \text{Distinctiveness} & \text{Variety}\end{array}}{\text{Price}}$$

Consumer Benefits

Shoppers of food products make purchase decisions based upon analytical and/or emotional processes that result in acceptances or rejections of products. The process involves an evaluation of the perceived worth or benefits of a product in comparison to the price charged, which in turn influences the decision to purchase or reject the item. However, consumers respond in different ways. Consumers from different cultural, demographic, and lifestyle segments, possessing various wants and needs, will place different emphases on the benefits criteria, as well as the price. In a particular shopper's purchase situation some factors may be unimportant, while others are paramount. Only if the perceived benefits of the product or service exceed the price charged, and if the benefit-to-price comparison is more favorable than for other perceived purchase alternatives, is the shopper likely to purchase the item.

The process of marketing plays a vital role in heightening consumers' value perceptions. Indeed, heightening consumers' value perceptions is the basic challenge confronting food marketers in their efforts to improve consumer demand and to achieve profitable operations. Indeed, as all foods are competing against an ever-growing array of alternative products, each positioned to gain a "share of stomach," the contemporary marketing approach is ever more essential to improve the probabilities of success (Allen and Pierson, 1990).

In addition to consumers, food processors must also market to other customers; namely, manufacturers, which use initially processed industrial ingredient food products, as well as retailers and wholesalers; and each of these customers make value determinations based upon benefits as well as price.

Purchase decisions by business customers, in comparison with consumers, are generally thought to be more objective and analytical, and less emotional; yet, the concept of these buyers basing purchase decisions upon comparisons of benefits-to-price still applies. Once again the importance of each perceived benefit will change from customer to customer.

In general, retailer and wholesaler purchasing decisions take into account the following kinds of benefits in their purchase decisions:

$$\frac{\text{Retailer and}}{\text{Wholesaler}} \quad \begin{array}{c}\text{Value} \\ \text{Perceptions}\end{array} = \frac{\text{Benefits}}{\text{Price}} = \frac{\begin{array}{l}\underline{\text{Retailer/Wholesaler Benefits}} \\ \text{Shopper Acceptance} \\ \text{Quality and Consistency} \\ \text{Dependability of Supplier} \\ \text{Reliability of Transport and Logistics} \\ \text{Terms of Trade} \\ \text{Purchasing Efficiency} \\ \text{Competitive Advantage} \\ \text{Advertising and Promotional Support} \\ \text{In-Store Merchandising Support} \\ \text{Strategic Alliances/Partnerships} \\ \text{Continuity and Tradition of Relationships}\end{array}}{\text{Price}}$$

While it is clear that suppliers selling to retailers and wholesalers must be competitive on price, they must also be competitive on a whole series of benefits. A supplier is more likely to be successful with retailer and wholesaler customers to the extent that the marketing process has been applied and that the supplier fully understands the goals, wants, and needs of a targeted customer or group of customers and that the supplier's products and services match the benefit and price requirements of customers.

Explanation of some of the benefits from the retailer/wholesaler value relationship may be helpful. For example, shopper acceptance is critical to retailer and wholesaler purchase decisions. Ultimately shopper acceptance is a function of the factors comprising the consumer value relationship shown above. Purchasing efficiency refers to the difficulty or ease that a retail or wholesale customer encounters when working with a supplier's sales organization and its methods of doing business. Competitive advantage applies to those product or service attributes of suppliers' offerings that may translate to an enhanced market position for the buyer. Strategic alliances/partnerships to an increasing number of relationships in which buyers and sellers negotiate many aspects of what are often longer term agreements provide strategic opportunities for improved business operations to both parties. And finally continuity and tradition of relationships refers to what may be a subjective set of factors that tend to favor long-time suppliers.

In the same manner as consumers, retailers, and wholesalers base purchase decisions on value relationships, so do manufacturers. For example, food manufacturers often purchase processed industrial ingredient food products from initial processors. In such instances the manufacturers' value perceptions may include the following kinds of benefits:

$$\begin{array}{ll} & \text{Manufacturer Benefits} \\ \hline & \text{Retailer and Consumer Acceptance} \\ & \text{Adherence to Specifications} \\ & \text{R \& D Assistance} \\ & \text{Flexibility of Products and Services} \\ & \text{Stability/Volatility of Supply} \\ & \text{Terms of Trade} \\ & \text{Purchasing Efficiency} \\ \text{Manufacturer} & \text{Continuity and Tradition of Relationship} \\ \text{Value} \quad = \quad \dfrac{\text{Benefits}}{\text{Price}} \quad = \quad \dfrac{}{\text{Price}} \\ \text{Perceptions} \end{array}$$

In terms of purchasing ingredient food products, it is clear that manufacturers must be assured of retailer and consumer acceptance; i.e., that ingredients and resultant products meet the wants and needs of customers. In today's increasingly competitive food market, initial processors are frequently being called on by manufacturers to provide research and development assistance in the use of ingredient products. In the same vein, manufacturers increasingly require flexibility of products and services; for example, the tailoring of specifications of a particular ingredient for a specific use.

Consumer Trends in the Industrialized World

Consumers' wants and needs constitute the major driving force that orients and directs the entire food system: retailers, wholesalers, manufacturers, processors, agribusiness firms, and farms. Therefore, it is important to consider the key directions of change among consumers across the industrialized world. There are clear differences in degree to which these trends apply from one culture to another and even among consumer segments within a given culture. However, global marketers observe that the basic directions of change are remarkably similar, that there is much more in common among consumers than differences. Analyses of marketplace initiatives and consumer buying behaviors across many industrialized nations strongly suggest that the non-price consumer food purchase criteria discussed above translate to the following five major directions of change impacting food marketing and distribution:

- Greater Convenience
- Higher Quality
- More Variety and Excitement
- Heightened Concern for Nutrition, Safety, and Health
- Growing Sensitivity to Environmental and Social Issues

Each of these major consumer directions of change is discussed to indicate how market-driven food and agricultural organizations are responding to marketplace opportunities.

Expanding the Concept of Convenience

Greater convenience in food is a major direction in food marketing. Changing lifestyles, changing demographics, and the development of new technologies are combining to accelerate this long-term trend. On the surface, the concept of convenience seems simple enough, but from a food marketing perspective it has many dimensions. Convenience as an overall concept can be viewed as those factors that increase comfort or make work less difficult. In the context of food, it is having food that is wanted, when it is wanted, with relatively little effort.

The concept of convenience in food is often limited to aspects of meal preparation. In fact, the relationship between food and convenience is much broader. Consider the following dimensions of food convenience:

Meal planning ranges from impulsive decisions at fast food restaurants and at food courts in shopping malls, to menu planning for a day, a week or longer. In the supermarket, merchandising of products, aisle directories, computer-driven directories, meal planners, and recipe services all contribute to making the process more convenient for shoppers as do food-oriented print and electronic media including the advertisements, restaurant menus, and even the photography on product packaging.

Food shopping and purchasing activities also are associated with convenience. A shopper familiar with a given food store finds it easier to shop than in a store that is new to his/her experience. Many shoppers will react negatively to even minor changes in store layout. The convenience or neighborhood store is a format that has developed primarily to facilitate the shopping of time-pressured consumers. A paradox is associated with the "one-stop shopping" concept of large supermarkets that applies to virtually all consumables. As stores get larger to accommodate the one-stop concept, they reach a size at which the time element of convenience diminishes. Thus, it is not surprising that drive-through windows and home-delivery food services are growing forms of food purchasing.

In-home handling and storage of food can be highly inconvenient. Consider the following concerns: Can food be rushed to the refrigerator or freezer before it warms or melts? Can room be found in the freezer? In which cupboard should the item go? Do vegetables belong under the sink or in the refrigerator, or where? How long will specific items keep? These kinds of issues and decisions impact virtually all households.

Food preparation and serving, while enjoyable for some people under some circumstances, nonetheless remain formidable for most people most of the time. Furthermore, over time, society in general is choosing not to develop the understanding and skill of food preparation common to previous generations. It appears that the less cooking consumers do, the less they will be able to do, thus heightening the essential inconvenience of food.

Even the act of eating is associated with convenience. Increasingly, the demands of busy lifestyles call for convenience of consumption. Traditional sit-down, home and restaurant environments are giving way to "meals on the go," such as eating in cars, "grazing" in airports, and snacking on planes. Under such conditions foods are often consumed with one hand, thus the need for "finger foods." The need for convenient consumption will continue to alter the forms of foods we eat.

Mealtime cleanup has always been considered drudgery and remains so even in the era of the dishwasher. Thus, innovation with respect to meal and food cleanup is likely to be responsive to consumers' convenience needs.

Quality Defined: Shoppers' Perceptions

Over time quality standards of food products are rising, although individuals' interpretations of quality may differ greatly. The acceptability of food varies from person to person, and for the same person, quality standards may also vary with circumstances. However, we live in an era in which expectations of eating quality are steadily rising.

In many instances quality serves as an all-encompassing concept relating to consumer perceptions of appearance, taste, freshness, nutrition, safety, consistency, and understanding. How shoppers assess food quality is a question of importance to food marketers. In general, shopper perceptions of food quality are based upon the following factors:

- *Appearance.* It is often noted by food marketers that "people buy with their eyes." Thus, food must first and foremost be visually attractive in supermarkets, as well as at the point of ultimate consumption. Although food appearance is highly subjective, to be regarded as "appealing," foods must increasingly appear visually perfect. As the standards for consumer acceptance continue to rise, increasingly there is a limited market for products that are even slightly lacking in appearance, though they may be totally acceptable in taste, nutrition, and other respects.
- *Taste.* Food marketers frequently rank taste as the most important consumer food attribute. A truly excellent tasting product is often the key to success, and mediocre taste is a likely reason for failure. Superior taste may overshadow shortcomings with respect to other product attributes. For example, many consumers routinely deviate from diets in favor of great-tasting foods, despite their high calories or poor nutritional value. In this indulgent society, when foods taste good enough, they may be eaten regardless of their nutritional shortcomings. For most people, taste and flavor is paramount.
- *Freshness.* Strong shopper demand for "fresh" foods is one of the most important trends impacting food sales of the past decade. It is most conspicuously observed in the success of supermarkets' merchandising of perishable sections. For many shoppers, freshness, or the appearance of it, is essentially synonymous with many concepts of quality. The perception of freshness is generally associated with bulk or unpackaged fruit and vegetables, chilled foods, hot prepared foods, and often the absence of manufacturers' branded packaged items. Fresh products apparently convey the image of better taste, wholesomeness, and good nutrition.
- *Nutrition.* For a growing number of consumers, the linkage between food quality and its nutritional content is essential. Clearly, nutrition is a quality factor in the minds of many shoppers. However, because nutrition and health have such important marketing implications, the topic is dealt with in a separate section that follows.
- *Safety.* Consumers continue to have heightened expectations for the wholesomeness and safety of all foods. Despite the fact that food safety standards have improved over time in most developed nations, consumers often respond with alarm to incidents of food-borne safety problems whenever they occur. Public expectations appear to approach perfection with respect to food safety and wholesomeness. This poses major challenges for food and agricultural marketers.

Although product safety and quality may be viewed as separate issues, they are in fact closely related. Food safety problems are clearly quality problems. Indeed, as a number of operators have indicated, serious quality problems exist long before safety problems occur. Largely because of the interrelationship between safety and quality, industry responses to the quality challenge closely parallel those of safety assurance.

- *Consistency.* That consistency of products and services is essential for successful food and agricultural operations is beyond debate. However, achieving the highest levels of consistency is not necessarily equated with the highest levels of product quality. Thus, the goal should be to achieve consistency of the targeted quality level.

- *Understanding.* Consumers' perceptions of food products are closely linked to their knowledge and understanding of the products. In the absence of understanding, information is needed. Positive product attributes that are not understood may not be appreciated as benefits. New food products run especially high risks of failure due to lack of consumer understanding. For example, Sous Vide processing, irradiated foods, vacuum packaging, and refrigerated displays of traditional hot foods with altered or unfamiliar appearances create obvious marketing challenges in terms of achieving consumer understanding. Though some consumers are accepting with respect to different or unfamiliar food appearances, many are reluctant to experiment, particularly with items perceived as expensive. They are concerned that the food may not taste good or perhaps that it is not safe or wholesome. In such a situation most marketers are reluctant to shoulder the expense and risks of attempting to inform consumers; however, meeting the challenge of consumer understanding by providing needed information may be necessary for the establishment of many significant breakthrough products based upon new or improved technologies.

Since quality is such a broad concept with many interpretations among consumers, it follows that food marketers, from farmers to retailers, must respond. Over the long-term, shoppers will pay a premium when and if they understand or appreciate quality enhancements and when the promise of better quality is fulfilled.

Another indication of marketing opportunities associated with quality derives from the following combination of situations: First, as has been indicated above, consumers are generally desirous of higher quality foods, and second, among consumers at large, skills, understanding, and experience with cooking and the creation of high-quality foods is diminishing. This combination of consumer desires and capabilities suggests that future sources for high-quality foods will increasingly be food marketers rather than consumers' kitchens.

Variety and Excitement: Creating Consumer Interest

Growth strategies of food marketers are increasingly focusing on variety and excitement for shopper appeal. Food processors are proliferating products with new and different tastes and flavors. At the same time, retailers' merchandising approaches are emphasizing more interesting and pleasurable shopping experiences.

Retail merchandising encompasses a wide range of strategic and tactical decisions, as well as the many activities carried out in the process of stimulating and promoting sales. It includes decisions concerning store location, layout, design, decor, and display techniques; in essence, the entire range of factors that combine to create the desired atmosphere. Merchandising also addresses decisions and activities relating to customer service levels, choices of product availability, product quality levels, and the breadth and depth of variety choices as well as decisions and activities relating to pricing and promotion.

To be sure, traditional products and merchandising abound, yet the direction of change is toward an ever-greater diversity of products and food store settings. The food industry has evolved beyond the situation when supermarkets were conventional with only minor differences and when the majority of new products were simply uninspired line extensions characterized by new colors, sizes and flavors or other kinds of marginal improvements. Today, food companies attempt to capture sales and market share by appealing to a consuming public that is more open to significant changes in foods. Travel, growing ethnicity, higher levels of education, and a growing diversity of restaurant food experiences are all causal factors associated with an interest in new foods. Consumer interest in food is based upon factors such as appearance, taste, understanding, a sense of adventure, and desires for unique or exotic dining experiences.

Many changes in food consumption patterns are stimulated by newly available ingredients. Consider the vast expansion in supermarket offerings of fresh fruits and vegetables, cheeses, and bakery products. The very existence of greater ethnicity in many developed nations is a major stimulus for new food concepts. Marketing opportunities associated with consumers' growing interests in variety and excitement are substantial in this dynamic atmosphere. In fact, it may seem that opportunities for new types of food, as well as new tastes and flavors, are virtually limitless; however, marketplace success for new foods is achieved by striking a balance between "newness" and sufficient consumer familiarity and understanding.

Nutrition, Safety, and Health: Issues of Increasing Importance

The past decade has been an era of growing consumer concern and awareness with respect to diet and health. What was thought by many in the 1970s to be a fad became a mainstream development in the 1980s and seems certain to continue as a series of ongoing, complex, and important issues for marketers in the years ahead.

Nutrition and health are vitally important to consumers and thus to the food and agricultural industries. As scientific and medical evidence substantiates the influence of diet on health, many people want to know and better understand how they should respond. Likewise, leaders in the food system from agricultural producers to processors and manufacturers, as well as wholesalers and retailers, are seeking answers to new and almost always challenging situations.

Industry responses are complicated by an often confusing range of human behavior. Today many people accept, and some are acting upon, half-truths and myths, and many questions exist in the minds of consumers:

- Does better nutrition mean less calories, fat, cholesterol, salt, sugar, meat, etc.?
- Will better nutrition result from more fiber, calcium, fish, poultry, etc.?
- Is natural better than conventionally produced meat and poultry?
- Are fresh foods superior to processed or preserved foods?
- Are organic fruits and vegetables better than commercially produced fruits and vegetables?
- Are governmental food safety standards and monitoring procedures adequate to ensure healthful and safe foods?

Despite nutrition and safety concerns, many consumers are indulgent with respect to food consumption; that is, compensatory behavior is common. For instance, a diet cola with the meal may be followed by a high-calorie dessert. Furthermore, consumers do not behave in similar ways. In increasingly segmented societies, there are groups of consumers possessing various attitudes and behaviors toward food safety and nutrition. However, it is important to note that the largest groups of consumers are comprised of those whose attitudes and behaviors are demanding higher standards of safety, as well as greater nutritional merits of foods. On the whole, they possess higher levels of education and affluence, and their influence on attitudes of other consumers is substantial. These are the thought leaders. Thus, it is expected that the long-term direction of change is for continued and increasing awareness of nutrition and the effects of specific foods and diets on not only certain health problems, such as cancer and heart disease, but also on the long-term overall health of individuals. Indeed growing numbers of people are likely to consider foods, diets, and lifestyles more seriously in terms of preventing health problems and prolonging healthful and active lives.

Consumers' changing food behaviors can negatively or positively impact food marketers. Growth rates of some products may slow and may reach saturation levels at less than projected volumes; indeed, some markets may diminish. On the positive side, however, increasing awareness and new behaviors driven by concerns for nutrition and safety have created opportunities for affirmative and innovative marketers.

Increasingly, consumers desire positive responses from the food industry to their emerging nutrition and food safety wants, and during the past decade food products emphasizing these attributes have proliferated. As new information regarding food and health comes to light, opportunities for improved products will continue. Moreover, it seems likely that the portion of the population concerned about nutrition and safety will grow, thereby expanding the market for nutritionally responsive foods. Food marketers have benefited by communicating positive nutritional attributes of new and existing products and will continue to do so. Lower fat and calorie products with excellent taste and convenience are increasingly popular.

Environmental and Social Issues: The New Imperatives

A number of environmental and social concerns have become mainstream in society, and food marketers must be aware of and responsive to them. The following kinds of issues are increasingly important on the food marketing landscape: waste disposal, packaging, pollution, animal welfare and rights, worker safety and welfare, roles of advertising, forms of competition, prices and profits, and the wise use of energy. "Win-win" responses meeting societies'; as well as consumers' and marketers' long-term interests must be sought. In many instances practices must change; however, in other cases marketers must mount effective public relations efforts in order to communicate how in fact their products and services serve the interests of the environment or are responsive to other social issues.

Consumers increasingly realize that there are growing interdependencies in the world, that food purchase decisions have short- and long-term influences on the environment and society. Shoppers express the following concerns: that food packaging may cause waste disposal and other environmental problems; that the origin of meat and other animal products may be associated with the inhumane treatment of animals; that food production and its distribution may involve the imprudent use of scarce resources; that the processing of food products may involve worker exploitation or unacceptable risks of injury; and that some advertisements may foster undesirable consumer behavior or perpetuate negative stereotypes. In fact, many shoppers today are increasingly sensitive to these kinds of environmental and social factors in their daily purchase decisions. The point is that marketers, perhaps more than ever before, face a volatile set of potential environmental and social issues that may substantially impact customer acceptance of their products, programs, and services. Understandably, marketers are often frustrated because many of these issues can be brought to light by events far removed from and beyond the control of the impacted marketer. Nevertheless, successful marketers must deal with these situations in the best possible ways.

Food and agricultural marketers increasingly understand that growing consumer awareness of a particular social or environmental issue may have either positive or negative effects. Successful marketers need to constantly search for signals of change in the marketplace and to respond in ways that either heighten positive affects or reduce negative impacts. The best approaches ought not be cynical, but rather based upon sound analysis. Of course, in some situations, responses producing positive outcomes in the short run are not to be found; thus a damage control approach is necessary. In the face of limited knowledge there is often great difficulty in knowing what response is environmentally or socially sound. Paradoxically, tomorrow's answers based upon new information may be different than today's.

Illustrations of Innovation and Entrepreneurship

Many companies are expanding their international businesses. Experiences of two large organizations, the Kellogg Company and CPC International, are highlighted here. These experiences serve to illustrate the implementation of the marketing approach, including a range of innovative and entrepreneurial initiatives.

The Kellogg Company

The Kellogg Company is a $5.8 billion producer of ready-to-eat cereal products, the largest cereal producer in the world. Kellogg also markets frozen pies and waffles, toaster pastries, cereal bars, and other convenience foods. Kellogg products are manufactured in 17 countries and distributed in over 150 countries worldwide and have an average market share of 44 percent. In 1991, over 50 percent of its cereal volume was sold in international markets. Many of the product formulations sold internationally are identical, but packaging reflects local language and cultural differences (Kellogg Co., 1991). A prominent food industry expert believes that cereal is a food that can be easily translated to international markets (Otto, 1992). Kellogg believes consumers' needs with respect to nutrition, taste, and convenience are the driving forces for the increased acceptance of its products and that these are worldwide directions of change in terms of consumers' wants and needs. Nutritional awareness among consumers worldwide is an important aspect of Kellogg's strategy for success. The company has responded by conducting nutritional seminars for health professionals in six countries in Latin America. A partnership was created with the French government to encourage consumption of breakfast. Cancer awareness programs were established in Canada. A cooperative program with the American Academy of Pediatrics was formed to promote healthy lifestyles for children. Kellogg sponsors the U.S. Olympic team and athletic events to reaffirm its tie to fitness and healthy living.

The company believes its greatest growth potential is in new and developing markets in Europe, Asia, and Latin America, yet Kellogg projects strong potential in its relatively highly developed markets in the United States, Canada, the U.K., and Australia (Gibson, 1991).

Virtually all Kellogg brands grew in market share in 1991, reversing a decline in sales in 1990. In 1990, strong sales in international markets offset a 4 percent decline in the highly competitive U.S. market. Worldwide coverage is a strategy that Kellogg believes can balance uneven performance of this kind. Investments are being made in all six cereal plants in North America to assure long-term production needs and to improve effectiveness. Developments are being made in food service marketing, inventory control, and customer service programs. In addition, these procedures, practices, and technologies are being applied to new and lesser developed markets worldwide.

Fifty years experience in Europe, especially in the U.K., has been the basis for a rapidly expanding cereal business in Western Continental Europe. Industry observers are convinced this base of experience gives Kellogg a vital competitive edge over other major firms attempting to develop markets for their products. Kellogg already has brand presence in Poland, Hungary, and Czechoslovakia and expects to grow the business in Eastern Europe. There are five technologically advanced cereal plants throughout Europe, with additional investments being made in them.

This market is expected by management to one day become Kellogg's largest operating division due principally to its immense population and the strong potential Kellogg sees for product acceptance. Production facilities exist in Australia (since 1924), Japan, South Korea, and South Africa. Rapid growth in sales has taken place in South Korea, Hong Kong, Singapore, and Malaysia. The Asian operations team was formed in Sydney in 1924 and is now the basis for coordinating the development of emerging markets in Thailand, Indonesia, the Philippines, Taiwan, Pakistan, Sri Lanka, and Bangladesh. The Sydney plant was recently expanded ($150 million), following an analysis of possibly establishing a plant elsewhere. Factors contributing to the decision to expand in Australia included strong trading links to build upon, an abundance of cereal grains, and talented personnel to maintain high standards of production and new product development (Messenger, 1991).

After years of preparation, Kellogg has achieved majority ownership of a cereal business in India. A new plant constructed in Bombay will be in production in 1994. Kellogg projects India's middle class to number 300 million by the year 2000 and believes these consumers will be receptive to its quality convenient products. The China market is under active exploration for business development.

Six plants are located in Mexico, Guatemala, Colombia, Venezuela, Brazil, and Argentina. Sales and profits are strong and the outlook promising. Growing consumer interest in diet and health, especially dietary fiber, is expected to fuel sales. Kellogg expects that greater political and economic stability and lowering of trade barriers will open new markets and permit synergies in manufacturing and marketing systems.

Kellogg's goal is to double its worldwide consumer base in the 1990s. A major investor service assesses Kellogg's international performance as outstanding with profit growth averaging 30 percent over the past four years on sales gains in the range of 7-10 percent. Foreign sales comprise over half the total poundage and contribute about 40 percent of total profits. Continued performance of this magnitude is projected. Profit margins for international operations are lower than domestic, but it is expected that as the scale of international operations and efficiencies increase, margins will improve (Grant, 1991). It is clear that viable international trade is a major component of Kellogg's overall corporate strategy.

CPC International

CPC International has operations in 51 countries in North America, Europe, Latin America, the Mideast, Africa, and Asia. Consumer food businesses account for 80 percent of the $6 billion in annual sales and corn refining businesses constitute the remaining 20 percent (CPC International, Inc., 1991). CPC has acquired over 30 companies in the past five years and has intentions of pursuing additional acquisitions worldwide to accelerate progress in implementing its strategic international plans (Kuhn, 1992). The company has been able to accomplish successful sales and profitability in the global marketplace through its commitment to developing a combination of strengths: leading brands, worldwide coverage with these brands, a formal global strategy, and experienced local management.

With $6 billion in sales, CPC is by no means among the largest of food firms operating worldwide. Despite its modest size, CPC is effective due to an organizational structure that is self-described as lean, agile, and flexible, accomplished in part by establishing strong local management. It is apparent that CPC believes it is advantageous to be of sufficient size to balance the varying profit performances in each of several major operating areas of the world. For instance, the weaknesses in European currencies negatively affected profits in early 1992, while in North America a major acquisition increased sales and profits significantly. Moreover, while the sale of certain products such as mayonnaise and corn oil were soft, other products such as peanut butter and pasta experienced strong sales and profits. Operating income in Latin American was particularly strong in early 1992, and the outlook is promising due in major part to CPC's over sixty years of experience in learning how to manage effectively in these volatile markets.

CPC International's brands are either number one or two in their categories.

This has become important in an era of rapid consolidation of wholesalers and retailers whereby some retailers are exercising their enhanced power to develop their own brands and/or to pressure manufacturers of branded products to accept costly selling terms. Only strong manufacturer brands can offset these trends. Brand strength will be an important factor in the unified European market as CPC strives to gain distribution of items in every country. Brand strength is instrumental in achieving worldwide coverage with products and the expansion of business activities across national boundaries with speed and without duplication of effort. The Knorr brand, for example, is effectively marketed in an integrated manner across several nations from a major central food plant in France. Brands must be strong, make a sound geographical fit, and be appropriate for the transfer of technological and management strengths of the company.

The role brands play in CPC's success is reflected in still other ways. Large numbers of people are immigrating to the U.S. from Mexico, Puerto Rico, Colombia, and Brazil where they have been loyal consumers of such CPC brands as Knorr and Mazola. This has contributed to the remarkable 30 percent increase of CPC brands in the U.S. among Hispanics over the past few years. Moreover, CPC continues to develop the skills necessary to address culturally different consumer segments in the United States, which further enhances its capability to build strong markets.

Establishing a focus on a limited number of product categories enables CPC to achieve strong product presence in selected food categories. It also facilitates the development of special skills and experience to ensure success. However, within a brand such as Knorr there are as many as 2,000 different products and line extensions. Worldwide product groups consist of sauces, soups, and bouillon; mayonnaise and corn oil; starches, syrups and desserts; bakery and pasta; and bread spreads. Regional product groups consist of desserts in Europe and corn starch products in Latin America. Sales growth of these products over the past five years has generally been above 15 percent annually.

Responses to consumers' growing interest in nutrition are reflected in the development of products such as cholesterol-free mayonnaise and whole-grain breads. CPC intends to introduce a fat-free mayonnaise when a product of acceptable taste can be developed. Mazola brand corn oil will continue to be positioned as an important part of a healthy diet, and the Mueller pasta line includes cholesterol-free egg noodles. Many other nutritionally oriented products are being developed (Food Institute, 1990).

CPC has developed a line of products designed for convenient preparation in microwave ovens. Aseptically packaged products are also meeting with success among consumers seeking products in convenient forms. These products are gaining acceptance in such diverse locations as Japan, Mexico and Western European Countries.

CPC has created the Cooperative Management Group (CMG) to leverage the corn refining business experience and expertise through joint ventures, licensing, and technical agreements. CMG manages 17 ventures in 15 countries.

It is widely acknowledged that CPC is a model food company in its approach to pursuing worldwide opportunities. A basic strategy is to achieve double-digit growth in sales and earnings by quick response to market opportunities, including the acquisition of food businesses that contribute financial and marketing strengths. The firm succeeds because of its ability to understand the intricacies of each market in which it operates and its willingness to pursue strategies with persistence and highly skilled personnel. Businesses are managed by people attuned to consumers, public policies, cultures, and eating habits of consumers in the local markets. This encourages quick response to local opportunities and avoids mistakes common to some companies that have more centralized decision making.

Summary and Conclusion

The experiences of Kellogg and CPC International illustrate the importance of non-price factors to success in international food marketing. To be sure, however, attention is paid by both firms to achieving cost-reduction, production and marketing efficiencies, and economics of scale that enable them to offer competitive prices while they simultaneously add value to food commodities and products. Price, to be sure, is vitally important, as price appeals to consumers are prevalent in virtually every market.

Nevertheless, both companies are successful mainly due to satisfying the expectations of consumers with products that incorporate in superior ways the non-price factors explained in this chapter. The products marketed do indeed reflect the five major directions of change impacting food marketing and distribution:

- Greater Convenience
- Higher Quality
- More Variety and Excitement
- Heightened Concern for Nutrition, Safety, and Health
- Growing Sensitivity to Environmental and Social Issues

Clearly, the argument is made with these brief case histories that non-price factors are essential to success in international trade. Other examples could be cited to make the case that the success of food companies of varying sizes and widely differing product categories are also related to how effectively non-price factors are implemented.

Finally, it is important to comment again on innovation and entrepreneurship as factors associated with competitiveness in international trade. These are the traits of personnel throughout organizations that enable companies to successfully implement the strategies and tactics of effective marketing. Kellogg and CPC International exhibit a number of prerequisites for successful international trade: Managers throughout these organizations are highly competent due to extensive experience in international activities over an extended period of time. They are guided by senior executives who themselves have had experiences in various countries and where they have come to understand the importance of such qualities as patience and persistence. Moreover, they realize that many international ventures run the risk of slow growth, delayed profitability, or outright failure, yet they accept these realities of international business. Many of these managers are able to conduct business in two or more languages and all have a global orientation and outlook that is reflected in corporate mission statements and strategic plans.

Innovativeness and entrepreneurial traits will be greatly needed in a rapidly changing global marketplace for food. The following kinds of issues will challenge executives as they seek to develop winning strategies:

- How can food brands be tailored to local markets; and yet be translated to several nations promoting product traits that appeal to multiple consumer segments?
- Conversely, can certain products become competitively secure in limited markets because of uniquely stable local conditions?
- How might businesses in various geographical markets be used as hedges against periodic or cyclical problems?
- Can corporate experiences in targeting ethnic markets in a company's home country be a significant learning experience and provide the basis for conducting business abroad?
- What is the appropriate response to an emerging situation where food manufacturers' brands are increasingly challenged by large retail firms bent on marketing their controlled labels across large geographic areas?
- How can U.S. food companies reconcile different cultural and ethical practices in countries where they wish to establish new businesses; i.e., illegal payments to government officials and organized crime?
- How can global business skills be developed at accelerated rates to create a larger cadre of middle managers in organizations to support the growing demand?
- How can companies deal with the reality of relatively high costs of international operations especially in the following areas: staffing and training, extended time frames required to achieve profitable returns, and the costs associated with addressing new and different complexities of international business enterprise?

How industry executives respond to these and other challenging issues will determine the extent to which U.S. food companies achieve profitable growth in the global marketplace.

References

Allen, J. W., and T. R. Pierson. "Fresh Prepared Foods: Major Opportunity for Food Industry Growth," *Food Industry Institute* (August 1990): 1-13.

CPC International Inc. *Annual Report* Englewood Cliffs, N.J.:CPC International, Inc., 1991

Food Institute Report. *The World According to CPC International*, Fair Lawn, N.J.: (March 3, 1990): 3.

Gibson, R. "Kellogg's Next Chairman Langbo Plans to Serve Cereal to More Foreign Markets." *Wall Street Journal* (October 21, 1991): B2.

Grant, S.E. Kellogg Company, *Value Line* February 22, 1991.

Kellogg Company. *Annual Report* Battlecreek, MI: Kellogg, 1991.

Kuhn, M.E. "First Quarter Ups and Downs for CPC." *Food Business* (May 18, 1992): 10.

Messenger, B. "Australia: The Bridge to the Pacific Rim." *Food Business* (September 9, 1991): 29.

―――. "Editor's Notes." *Food Business* (November 4, 1991): 6.

Otto, A. "The Top 15 Multinational Food Companies." *Food Business* (July 6, 1992): 36-8.

Pierson, T.R., and J.W. Allen. "Food Marketing and Distribution: Implications for Agriculture." Special Report: Status and Potential of Michigan Agriculture — Phase II, Michigan State University, Agricultural Experiment Station, East Lansing, MI. (July 1992): 2-15.

5

Importance of Non-price Factors to Competitiveness in International Food Trade

Michael R. Reed

Introduction

The two previous chapters stress that price may not be as important in determining international competitiveness as most economists believe. One of the keys to success is to differentiate one's products so that they can compete on the basis of quality or other non-price characteristic. This chapter investigates the effects of some non-price factors on U.S. agricultural exports by level of processing.

There is much talk these days about the level of processing involved in most agricultural exports (Evans, 1990). Increased exports of processed food products will not only stimulate farm income, but also provide manufacturing jobs (Schulter and Edmondson, 1989). The General Accounting Office (GAO) recently charged that the U.S. Department of Agriculture (USDA) must rethink its priorities if it is to help increase the exports of processed foods. In particular, GAO insists that the USDA must engage in more strategic marketing in cooperation with the private sector.

The GAO is simply reiterating what it hears from the popular press: The United States must become more competitive in processed food markets. It is unclear what the popular press means, but this chapter will address some of the relationships that could be inferred from these statements. Specifically, this chapter investigates factors that may explain the level of processing involved in United States agricultural exports. These factors are broken down into economic and noneconomic factors that explain market penetration by U.S. firms.[1] The premise is that the export process involves an extension of the firm beyond the comfort of its home country base. This not only involves costs but also risks. The firm must expect significant returns over domestic sales in order to compensate for these additional costs and risks. It seems natural that firms will first seek out those countries with cultural, political, and legal systems resembling those of the United States. Thus, it is hypothesized that these noneconomic factors will play an important role in determining the processing level of U.S. agricultural exports.

U.S. Agricultural Exports by Degree of Processing

The Foreign Agricultural Service (FAS) of the USDA classifies agricultural exports based on how close they are to their final consumer form: Bulk (which are free from processing), intermediate (which are principally semi-processed), and consumer-oriented (products which require little additional processing).[2] In 1990 the United States exported 51.0 percent of its agricultural products in bulk form, 22.4 percent in intermediate form, and 26.6 percent in consumer-oriented form (Figure 5.1). This compares with 34.4 percent in bulk form, 23.3 percent in intermediate form, and 42.3 percent in consumer-oriented form for total world agricultural trade. The world figures exclude intra-EC trade. The United States seems less competitive in markets for consumer-oriented food products (if market share is the measure of competitiveness).

When analyzing the form of U.S. exports, I prefer to subtract exports of unprocessed consumer-oriented foods, such as fresh fruits, eggs, fresh vegetables, and tree nuts from the consumer-oriented category to obtain four

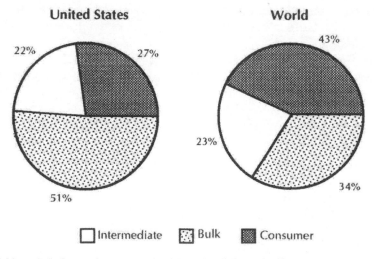

FIGURE 5.1 The United States and Global Food Trade by Degree of Processing.

groups closely related to the four economies of agriculture proposed by Abbott and Bredahl: bulk, high-value unprocessed, intermediate, and highly processed. With this breakdown, the United States exported 51.0 percent of its agricultural products in bulk form, 7.9 percent in high-value bulk form, 22.4 percent in intermediate form, and 18.7 percent in highly processed form.

Figure 5.2 shows U.S. agricultural exports by these four categories from 1985 to 1991. There is a clear trend toward an increased percentage of high-valued unprocessed and highly processed foods, while the trend for bulk is clearly downward. In fact, U.S. consumer-oriented food exports increased by an annual rate of 24 percent between 1985 and 1990. This is very positive because world trade in consumer-oriented food products grew at an 11 percent annual rate between 1983 and 1990, compared to a 6 percent annual rate for intermediate products and a 1 percent annual rate for bulk products. Hence, as incomes across the world increase, there will likely be increased export opportunities for consumer-oriented food products.

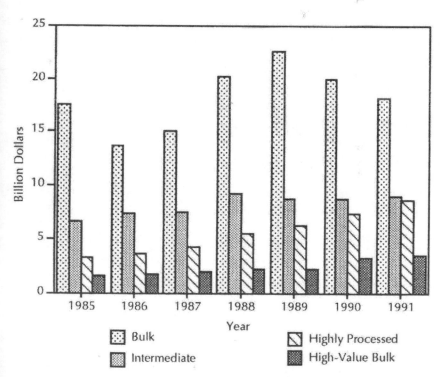

FIGURE 5.2 U.S. Agricultural Exports by Level of Processing, 1985-91 (Billion dollars).

Table 5.1 shows the leading per capita markets for U.S. agricultural exports by the four levels of processing. Hong Kong, Singapore, and Canada were the leading per capita destinations for highly processed U.S. food, each averaging imports of more than $25 per person for the 1987-89 period. The same three countries were also the leading per capita destinations for high-valued unprocessed U.S. exports (though the order for Singapore and Canada is reversed). The Netherlands was the leading per capita destination for semi-processed and low-value bulk exports from the United States.

Review of Relevant Concepts

Despite the rhetoric of the need for the United States to increase its exports of value-added foods, there has been little research aimed at understanding foreign markets for high-value-added foods. For instance, there has been no analysis to investigate why the United States has had such a low percentage of world trade in processed products or why the United States seems to be increasing its emphasis on high-value-added exports. One reason why no analysis has been performed might be the reluctance of agricultural and general economists to relax the assumptions of basic microeconomic theory. Neo-classical microeconomic theory basically postulates that three variables are important in demand systems: own-price, cross-prices, and income. In fairness to the profession, some consideration has been given to income distribution and population demographic and growth; however, these variables have been rarely integrated into conceptual or empirical analyses.

One positive step taken has been the incorporation of product quality and characteristics into demand analysis. In most instances, this incorporation is quite simple (e.g., two goods with one universally recognized superior in quality to the other), but it is certainly a step in the right direction. When it

TABLE 5.1 Leading Markets for U.S. Agricultural Exports by Level of Processing, 1987-89 (Average in $ per Person)

Highly Processed	High-value Unprocessed	Semi-processed	Low-value Bulk
Hong Kong $39.17	Hong Kong $23.97	Netherlands $57.79	Netherlands $62.94
Singapore $29.77	Canada $20.55	Canada $24.39	Israel $58.59
Canada $26.36	Singapore $11.34	Venezuela $14.52	Trinidad $46.57
Japan $15.48	Netherlands $4.86	Algeria $14.25	Japan $31.19
Panama $13.59	Japan $4.35	Iraq $13.92	Iraq $26.13

comes to explaining demand patterns for all but bulk commodities, product characteristics and differentiation seem to be the rule rather than the exception. Advertising expenditures and brand loyalty are very important variables in explaining processed food demand, but these have been ignored by agricultural economics. It seems that one must travel to the marketing literature to find insights into the demand for processed foods.

The marketing literature begins with the four Ps: price, product, place, and promotion. Given that this chapter concerns non-price factors and their effects on U.S. agricultural exports, I will spend little time discussing price. It should be noted, however, that many times price is less important than other product characteristics in consumer purchasing decisions. Porter (1990) argues that the most successful companies compete on the basis of product differentiation rather than price. The reason is that some company/country that will always beat your/the price, but it may not be able to match your/the quality if it is consistently upgraded. However, price can be an important factor in processed food exports because tariffs usually escalate as the level of processing increases.

In the U.S. domestic market, product form and quality/image play important roles. U.S. consumers are increasingly concerned with product ingredients, healthfulness, and safety of their food. Consumers in many international markets (e.g., Europe and Japan) are equally concerned. However, in those markets, product form plays a more vital role because consumers not only eat widely different foods, but also desire different portion sizes, label colors, packaging, and many other characteristics for American-style foods. A food product is really a bundle of characteristics, and the variation in the value placed on those characteristics is wider internationally than it is in the United States. A complaint that is commonly heard in Japan is that the portion size of most imported U.S. food products is too large, given consumption patterns and storage capacities of the typical Japanese household. These considerations are not as important if one is exporting a bulk or intermediate product because the final manufacturer is close to the consumer and can make sure the desired characteristics are in the final product. Meeting foreign consumer needs is more difficult if the company is exporting a highly processed product (or even a consumer-oriented product).

Successful food products generally have an established brand identity and are known for their quality. Advertising accounts for up to 12 percent of sales for some U.S. food companies (*Advertising Age*, 1991). This advertising is especially important when a new brand is introduced. A well-known U.S. brand is much more likely to be accepted by companies in foreign distribution systems. In many countries, food wholesalers and retailers are bombarded with proposals from food manufacturers. If a well-established U.S. food processor asks a foreign food company to consider handling its product, the foreign company is more likely to make a positive response if the brand has a good following in the United States. I recall speaking with the purchasing manager of a large specialty food retailer in Taiwan about some particular U.S. food products. She specifically asked which U.S. food retailers carried the products and how they were selling. However, U.S. brand identity can go only so far in helping access international markets. The ultimate consumer in these countries must become aware of the product if market entry is to be successful. Thus, advertising and promotion strategies must be developed specifically for the target market.

Product placement is critically important for marketing food products internationally. Food distribution systems vary markedly between countries. My impression is that the food distribution system in the United States is quite similar to that in Australia, Britain, and Canada — the firms are quite open to imports and are not averse to trying new products. This is not true in most Asian and some European countries. Tradition plays a larger role and firms are reluctant to establish new alliances, especially if it means weakening old alliances. This tradition-bound marketing system is changing in Asia and Europe, but quite slowly.

Some Asian food distribution systems are also fraught with many layers and a complex array of marketing firms, particularly in Japan. Part of the reason for this is that the high cost of space in these countries (office space rents for 50 percent more in Tokyo than in New York City) increases the cost of storage and dictates the just-in-time delivery system for food and other products (reducing inventory levels). It is not unusual for Tokyo wholesalers to make two deliveries each day to retailers. The U.S. government alleges that some Asian governments encourage this complex distribution system to make it difficult for imports to enter the distribution system. These foreign governments counter that all food manufacturing firms face the same distribution system, but it is easier for a domestic firm to weave its way through a complex system than it is for a foreign firm.

Jeannet and Hennessey (1988) accumulate these considerations into three basic broad, noneconomic forces that specifically influence international marketing: cultural forces, political forces, and legal forces. These forces vary by country and must be considered by an exporter if an entry strategy is to succeed. These factors will impede U.S. agricultural exports of all kinds, but they are less difficult to overcome with unprocessed products than highly processed products.

Cultural Forces

Jeannet and Hennessey categorize four aspects of culture that are important in international marketing: religion, language, education, and family. These elements of culture affect not only food consumption patterns, but also distribution systems and business practices. Muslim values are quite different from Christian values, and this affects U.S. exporters. To sell products in Saudi

Arabia, exporters must recognize Muslim values and their implication for marketing strategies and the demand for product characteristics. The same exporter might be much more comfortable in Mexico, where Catholic values are strong.

Language differences can be formidable barriers to trade. Contacts between U.S. exporter and foreign importer are more difficult, even if both sides are familiar with a foreign language. Many phrases cannot be readily translated and even when translated many have a different meaning. Part of this difficulty in translation is that language reflects different thinking patterns. Therefore, even if initial contact can be made between buyer and seller, there will often be problems in translating ideas from one language to another. Most countries also either require that the food label be in their native language or that a native language sticker be attached to the English-language label. This increases costs and deters some U.S. companies from exporting.

Educational systems differ widely among countries. However, for explaining U.S. export patterns, the output of those systems is more important than their structure. Porter (1990) claims that the United States and United Kingdom are world leaders not only in advertising and promotion strategies, but also in products that rely on advertising to establish their images, such as processed foods. The basic premise of advertising is to use a mass medium to portray the product in a positive light. This assumes that there is a well-established mass media system in the country. Educational attainment in most areas of the developed world are such that advertisers can choose the communications medium — radio, television, newspapers. However, in many areas of the world, adult illiteracy rates are low and mass communication systems are not well developed. This will hinder any efforts to introduce differentiated products because consumers will not readily understand the superior attributes of the differentiated products.

The final aspect of culture is the role of the family. General attitudes about the importance of the family, household size, percentage of women in the work force, and other factors differ by society and can influence the prospects of marketing efforts by firms. Again, as these family values and characteristics vary in foreign countries from a U.S. reference, the more difficulty U.S. firms will have in exporting highly processed foods because they will have to adjust their marketing strategy or product.

Legal Forces

Differences in legal systems probably have a larger effect on direct foreign investment than on exporting. In that regard investment sanctioning, contract law, taxation, competition standards, and other legal regulations are much more important for a U.S. company owning a plant in a country, than a U.S. company trying to penetrate the market through exports. How the legal system handles/enforces agency agreements, product liability, and patents are relevant to potential exporters. Usually, it is much more difficult to terminate an oral or written contract between a U.S. exporter and foreign distributor outside the United States (Reynolds et al., 1990).

Political Forces

The political system, the stability of the government, and the potential for nationalization of foreign-owned assets are also important for U.S. multinational corporations. However, U.S. agricultural exporters, particularly of processed food products, will be less likely to invest time and money in penetrating a market if the country is distant politically from the United States. In addition, because of the politics of U.S. government programs, subsidized sales and marketing efforts for U.S. companies are more likely to be approved for countries that are politically "close" to the United States.

Operationalizing the Concepts

In general, all the factors mentioned in the previous section measure the costs of entering and maintaining a presence by a U.S. agricultural exporting firm. They are noneconomic variables in the sense that they are neither price nor income variables associated with the market in the particular country, though they are economic variables in the sense that they measure costs of exporting to the country. They are definitely factors that must be overcome by the firm. Regression analysis measures the extent that these noneconomic (social/cultural) and economic (price and income) variables have influenced U.S. agricultural exports by level of processing.

The dependent variables in the analysis are the value of U.S. exports of agricultural products by degree of processing. These data were obtained from the Foreign Agricultural Service (though the initial source is the U.S. Census Bureau). Four categories are used: bulk, high-value unprocessed, intermediate, and highly processed. Low-valued bulk and high-valued bulk are distinguished because high-valued bulk products are more consumer-oriented than low-valued bulk. Two dependent variables are formed from these data for the analysis: 1) the percentage of the value of U.S. agricultural exports accounted for by each processing category and 2) the value of per capita U.S. agricultural exports by processing category. Foreign supply effects in both models are ignored. Thus, U.S. exports, particularly in the second set of models, are assumed to be imperfect substitutes for domestically-produced agricultural goods.

Noneconomic Concepts

The combination of operationalizing the concepts discussed earlier and finding consistent data for the operationalized variables was frustrating. My only consolation was the comment of a colleague: "When you are the first to attempt to analyze an issue, the study will be less eloquent than when the issue is analyzed for the tenth time." Please note, however, that this research has many potential refinements and extensions.

The percentage of the population practicing a Western religion for a given country was used to measure how similar the country was to the United States (Catholic, Eastern Orthodox, Jewish, and Protestant faiths were classified as Western religions.) These data were obtained from the Central Intelligence Agency. The adult illiteracy rate was used to measure the educational attainment of the country. The percentage of the population living in urban areas was used to measure the country's family structure. More urban societies are more likely to have a U.S.-style family value system. Data on illiteracy rates and urbanization came from the World Bank.

The political proximity of the country toward the United States was measured by per capita receipt of U.S. economic development assistance (data from the U.S. Department of Commerce). This variable works fine as long as the country is less developed, but will not measure the political closeness of more developed countries. The number of American tourists (divided by the country's population) was used as a generalized cultural variable. American tourists naturally flock to countries that are similar to the United States. Numbers on American tourists came from the UNESCO. No variable was found to measure differences in legal systems.

Economic Concepts

Three sets of economic variables are used to explain the pattern of U.S. agricultural exports by level of processing. The first set measures income as per capita nominal gross domestic product (GDP). The second set of variables measures the ability to spend foreign exchange on imported agricultural products. Three variables are used: external debt, international reserves and the balance of payments (the total change in international reserves). Data on GDP and the international sector were obtained from the United Nations.

The third set of variables measures the level of tariffs by stage of processing. For developed markets, the primary-stage tariff level was the trade-weighted tariff on agricultural raw materials, while the processed-stage tariff level was the trade-weighted tariff for all food items.[3] The primary-stage tariff level for all other countries was for primary products, while the processed-stage tariff level was for manufactured products. All tariff data came from UNCTAD. The years of observation for each country varied from 1981 to 1988. These are the only price variables used in the analysis. Import unit values by level of processing would reflect a different mix of products, given the diversity of products in each processing category, rather than different prices between countries for homogenous agricultural products.

Other Data Considerations

All data on U.S. agricultural exports were averages of the 1987-89 period. Other data were generally from the 1987-89 period. Dairy products were excluded from the analysis because a large proportion of those products were given away through government programs (especially to less developed countries). In many instances before deleting dairy products, African countries would be importing 20-50 percent of their U.S. products in a highly processed form. It was felt that including dairy products would distort the results. Table 5.2 shows the leading markets for U.S. agricultural exports by level of processing as a percentage of total U.S. exports.

Results

Ordinary least squares (OLS) regression was used to obtain the coefficient estimates for both sets of regression analyses. Note that the dependent variables were in percentages terms for the first set (Tables 5.3 through 5.6) and per capita U.S. exports in the second set (Tables 5.7 through 5.10). Each table shows estimates from four common specifications. The first model specification uses only the noneconomic variables; the second uses only the economic variables; the last two are hybrid models using all the noneconomic variables, plus two economic variables.[4] Observation numbers vary by model specification because of data availability. Coefficients significantly different from zero at the 10 percent level are starred in the tables.

Models Explaining U.S. Export Percentages
by Processing Level

The noneconomic variables did a reasonable job of explaining variation in the dependent variables, given that the models were cross-sectional. In the purely

TABLE 5.2 Leading Markets for U.S. Agricultural Exports by Level of Processing, 1987-89 (Percentage of U.S Agricultural Exports)

Highly Processed	High-value Unprocessed	Semi-processed	Low-value Bulk
Nicaragua 98.2%	Canada 25.9%	Iran 100%	P.R. China 96.5%
Singapore 53.7%	Hong Kong 25.5%	Hungary 93.3%	Mali 92.2%
Hong Kong 41.7%	Sweden 23.6%.	Czech 89.0%	Zaire 89.8%
Panama 37.9%	Singapore 20.5%	Lesotho 81.9%	Papua-NG 87.3%
Sweden 37.6%	New Zealand 17.1%	Argentina 70.3%	USSR 86.6%

TABLE 5.3 Results of the Analysis for Percentage of the Value of U.S. Exports in Highly Processed Form

Variable	All Non-Economic	All Economic	Mixture 1	Mixture 2
Tourism	58.91*		34.23*	37.49*
	(18.07)		(17.28)	(16.04)
Literacy	.040		.021	-.057
	(.090)		(.100)	(.084)
Religion	-.060		-.105*	-.041
	(.040)		(.038)	(.039)
Urban	235		.185*	.071
	(.076)		(.079)	(.074)
USAid	-32.87		99.72	-32.60
	(28.12)		(147.40)	(23.16)
GDP		.708*	.641*	.252
		(.330)	(.346)	(.323)
Balance of Payments		16.82*		
		(9.22)		
Reserves		6.67*		5.26*
		(1.64)		(1.59)
External Debt		-4.04		
		(2.78)		
Manufacturing Tariff		-.020	-.094	
		(.103)	(.071)	
Primary Tariff		-.043		
		(.146)		
n	60	42	46	53
R²	40	59	58	58

*Coefficients significantly different from zero at the 10 percent level.

noneconomic specification, the religion and urbanization coefficients were significantly different from zero the most times (three out of four). High-valued exports (highly processed and high-valued unprocessed) went to countries with a larger proportion of urban consumers and people practicing non-Western religions (though the coefficient on religion was positive and significant for bulk exports). High-valued exports also tended to account for a higher percentage of U.S. exports to areas attracting more U.S. tourists per capita. The purely noneconomic models explained 21 percent to 62 percent of the variation in the dependent variables.

The GDP coefficient was significantly different from zero in three out of the four equations for the purely economic specification, but coefficients for external debt, the tariff on primary products, and the tariff on manufactured products were not significantly different from zero in any of the purely economic equations. Income level clearly accounts for much of the variation in

TABLE 5.4 Results of the Analysis for Percentage of the Value of U.S. Exports in High-Valued Unprocessed Form

Variable	All Non-Economic	All Economic	Mixture 1	Mixture 2
Tourism	41.95*		30.96*	26.60*
	(8.04)		(8.80)	(7.11)
Literacy	-.028		-.008	-.015
	(.040)		(.051)	(.037)
Religion	-.031*		-.036*	-.008
	(.018)		(.020)	(.018)
Urban	.123*		.078*	.031
	(.034)		(.040)	(.033)
USAid	-20.25		8.19	-16.76
	(12.51)		(75.19)	(10.27)
GDP		.727*	.512*	.512*
		(.193)	(.177)	(.143)
Balance of Payments		1.78 (5.36)		
Reserves		2.10*		1.50*
		(.96)		(.71)
External Debt		-1.62 (1.63)		
Manufacturing Tariff		.025 (.060)		
Primary Tariff		-.044	-.035	
		(.085)	(.053)	
n	60	42	46	52
R^2	62	64	70	74

*Coefficients significantly different from zero at the 10 percent level.

the processing level of U.S. agricultural products. Higher income countries tended to import more U.S. agricultural products in high-valued forms.

The purely economic specifications explained more variation than the purely noneconomic specifications, but they also had fewer observations. This could explain some of the difference in R^2s between the models. The only time when there was a large difference in R^2 was for highly processed products, where the economic specification explained 59 percent of the variation and the noneconomic specification explained 40 percent of the variation. The mixed models show the results of including both types of variables, though the specifications are skewed towards noneconomic variables. In mixture 1, GDP is included with the tariff rate that is closest to the particular processing level (the highly processed and intermediate specifications used the manufacturing tariff and the high-valued unprocessed and bulk used the tariff on primary products). Tariff rates were included because they were the only price-type variables in the

TABLE 5.5 Results of the Analysis for Percentage of the Value of U.S. Exports in Intermediate Form

Variable	All Non-Economic	All Economic	Mixture 1	Mixture 2
Tourism	-34.96 (37.79)		-6.34 (36.05)	-16.96 (38.90)
Literacy	.246 (.188)		.327 (.208)	.308 (.203)
Religion	.275* (.085)		.275* (.081)	.268* (.097)
Urban	-.001 (.159)		.108 (.164)	.034 (.181)
USAid	-85.64 (58.76)		-306.6 (307.7)	-80.38 (56.16)
GDP		.139 (.656)	-.285 (.723)	-.191 (.783)
Balance of Payments		-15.95 (18.17)		
Foreign Reserves		-4.17 (3.23)		-1.83 (3.89)
External Debt		3.09 (5.52)		
Manufacturing Tariff		.271 (.203)	.225 (.149)	
Primary Tariff		-.109 (.288)		
n	60	42	46	52
R^2	21	16	30	24

*Coefficients significantly different from zero at the 10 percent level.

analysis. In mixture 2, the two variables that had the most significant coefficients in the purely economic specification were included — GDP and international reserves.

In these mixed models, religion (and to some extent the number of American tourists, urbanization, and income levels) is important in explaining variation for most processing levels. The coefficient for religion was negative and significantly different from zero for both the high-valued equations, indicating that higher percentages of U.S. exports were highly valued in countries with non-Western religions. Coefficients on tariff variables and illiteracy were never significantly different from zero. The explanatory power of these mixed models was relatively close to the other models, though much more variation in intermediate product exports was explained by the mixed models.

Markets with a high percentage of highly processed agricultural imports from the United States tended to be similar to the United States (Table 5.2). They

TABLE 5.6 Results of the Analysis for Percentage of the Value of U.S. Exports in Bulk Form

Variable	All Non-Economic	All Economic	Mixture 1	Mixture 2
Tourism	-64.86 (41.04)		-58.93 (40.32)	-46.72 (41.56)
Literacy	-.261 (.205)		-.385 (.234)	-.237 (.217)
Religion	-.179* (.093)		-.122 (.091)	-.212* (.104)
Urban	-.361* (.172)		-.367* (.164)	-.144 (.193)
USAid	139.55* (63.82)		229.8 (344.3)	130.55* (60.00)
GDP		-1.52* (.77)	-.713 (.812)	-.558 (.837)
Balance of Payments		-2.42 (21.22)		
Foreign Reserves		-4.54 (3.78)		-4.73 (4.15)
External Debt		2.89 (6.45)		
Manufacturing Tariff		-.272 (.237)		
Primary Tariff		.204 (.337)	.015 (.242)	
n	60	42	46	52
R^2	24	30	30	29

*Coefficients significantly different from zero at the 10 percent level.

had higher income levels and high urbanization rates. However, they tended to practice non-Western religions. Obviously, one can see the influence of Hong Kong and Singapore in these numbers. Both are free ports, so duties are quite low, but few of their citizens practice Western religions.

Markets with a high percentage of bulk agricultural imports from the United States tended to be less similar to the United States (Table 5.6). They practiced non-Western religions, were less urbanized, and had lower incomes (though that coefficient was not significantly different from zero). It is interesting that religion is negatively related to all processing levels except intermediate products.

The results with respect to tariff rates were surprising. Part of the reason is that these tariffs are trade-weighted, so that prohibitive tariffs have a small weight in the final percentage. These tariffs also exclude non-tariff barriers. Frankly, many of the tariffs for developed countries were suspect. The trade-weighted tariff for raw agricultural products in the European Community was

TABLE 5.7 Results of the Analysis for U.S. Exports of Highly Processed Foods, Per Capita.

Variable	All Non-Economic	All Economic	Mixture 1	Mixture 2
Tourism	53.20* (8.59)		51.48* (9.90)	44.31* (6.74)
Literacy	-.023 (.043)		-.006 (.057)	-.039 (.034)
Religion	-.076* (.019)		-.097* (.022)	-.050* (.016)
Urban	.096* (.036)		.120* (.045)	.039 (.030)
USAid	-7.35 (13.37)		29.02 (84.44)	-6.98 (9.34)
GDP		.312 (.195)	-.132 (.198)	-.134 (.130)
Balance of Payments		14.50* (5.44)		
Foreign Reserves		4.13* (0.98)		2.37* (0.64)
External Debt		-1.45 (1.64)		
Manufacturing Tariff		.001 (.061)	-.073* (.041)	
Primary Tariff		-.026 (.086)		
n	60	43	47	53
R^2	61	52	67	73

*Coefficients significantly different from zero at the 10 percent level.

0.4 percent and for Japan 0.3 percent; the tariffs for all food items were 4.4 percent for the EC and 9.4 percent for Japan. Variable levies and minimum import price schemes must not be included in those calculations. Thus, these results in relation to tariff rates are questionable.

Models Explaining U.S. Per Capita Export by Processing Level

The purely noneconomic models did a consistently better job than the purely economic model in explaining per capita exports by level of processing. The per capita, noneconomic models performed better than the noneconomic models in percentage terms. The number of American tourists per capita (with a positive coefficient) and the religion of the country (with a negative coefficient) had significant coefficients in three of the four equations. U.S. development assistance and adult illiteracy were only significantly linked to export of bulk products.

TABLE 5.8 Results of the Analysis for U.S. Exports of High-Valued Unprocessed Foods, Per Capita

Variable	All Non-Economic	All Economic	Mixture 1	Mixture 2
Tourism	38.43* (4.34)		39.70* (5.10)	32.73* (2.56)
Literacy	-.006 (.022)		.006 (.030)	-.005 (.013)
Religion	-.037* (.010)		-.045* (.011)	-.022* (.006)
Urban	.045* (.018)		.054* (.023)	.018 (.012)
USAid	-4.41 (6.76)		-.518 (43.59)	-2.89 (3.70)
GDP		.265* (.126)	-.103 (.103)	.018* (.051)
Balance of Payments		6.55* (3.52)		
Foreign Reserves		1.37* (.63)		.568* (.254)
External Debt		-.770 (1.06)		
Manufacturing Tariff		.010 (.039)		
Primary Tariff		-.026 (.056)	-.064* (.030)	
n	60	43	47	53
R^2	70	39	75	86

*Coefficients significantly different from zero at the 10 percent level.

The per capita results using the purely economic models did not consistently explain more or less than the percentage results — R^2s were higher for high-value product models but lower for the others. The coefficient on GDP was significantly different from zero and positive in three equations, and at least one importing constraint coefficient (either for the balance of payments, international reserves, or external debt) was significantly different from zero in every equation, except for intermediate products. Coefficients on other economic variables were nonsignificant in all equations.

The income coefficients were generally larger for bulk and intermediate products in the purely economic models, which was mildly surprising. A one thousand dollar per capita increase in GDP increased bulk U.S. exports by $1.29. Coefficients on variables measuring importing constraints were also largest for bulk commodities.

The results of the mixed models were similar to the noneconomic models. Coefficients for noneconomic variables that were significantly different from

TABLE 5.9 Results of the Analysis for U.S. Exports of Intermediate Agricultural Products, Per Capita

Variable	All Non-Economic	All Economic	Mixture 1	Mixture 2
Tourism	30.81* (13.93)		25.81 (17.59)	29.84* (16.14)
Literacy	-.043 (.069)		-.053 (.101)	-.039 (.084)
Religion	-.023 (.031)		-.038 (.039)	-.031 (.040)
Urban	.074 (.058)		.062 (.080)	.083 (.075)
USAid	-9.56 (21.67)		22.95 (150.1)	-8.10 (23.31)
GDP		.817* (.385)	.097 (.353)	.176 (.325)
Balance of Payments		13.92 (10.77)		
Foreign Reserves		-.23 (1.92)		-1.63 (1.60)
External Debt		2.18 (3.25)		
Manufacturing Tariff		.008 (.120)	-.021 (.072)	
Primary Tariff		-.035 (.171)		
n	60	43	47	53
R	23	19	21	23

*Coefficients significantly different from zero at the 10 percent level.

zero in the purely noneconomic model changed little and their significance was similar. In contrast, coefficients for economic variables changed quite a lot. Many times the coefficient for GDP would change from positive in the purely economic specification, to negative in the mixed models. Coefficients on GDP fell in all models moving from the purely economic specifications to the mixed specifications. One interesting result is that the coefficient on tariff rates was significantly different from zero and negative in both high-value product equations.

Conclusions and Future Research

It is interesting that the regression models explain more variation in the high-valued export categories than the others. This indicates that there are more factors than these simple cultural and economic variables entering into the United States's pattern of bulk and intermediate product trade. Product

TABLE 5.10 Results of the Analysis for U.S. Exports of Bulk Agricultural Commodities, Per Capita

Variable	All Non-Economic	All Economic	Mixture 1	Mixture 2
Tourism	8.55 (18.39)		-3.45 (22.35)	3.73 (21.06)
Literacy	-.182* (.092)		-.223* (.129)	-.196* (.110)
Religion	-.084* (.041)		-.111* (.050)	-.090* (.052)
Urban	.087 (.077)		.071 (.102)	.075 (.098)
USAid	131.19* (28.61)		308.9 (191.0)	132.80* (30.41)
GDP		1.29* (.46)	.196 (.450)	.245 (.424)
Balance of Payments		29.35* (12.81)		
Foreign Reserves		.27 (2.28)		-1.57 (2.09)
External Debt		7.35* (3.87)		
Manufacturing Tariff		-.036 (.143)		
Primary Tariff		.038 (.203)	-.042 (.133)	
n	60	43	47	53
R²	44	33	28	44

*Coefficients significantly different from zero at the 10 percent level.

differentiation is higher for highly processed and high-valued unprocessed foods, so supply considerations in the importing country may be less important. The U.S. price for these products relative to domestic suppliers or other exporting countries may be more important for bulk and intermediate agricultural products. Another way of saying this is that Ricardian aspects of trade (i.e., factor endowments) may be much more important in explaining trade in bulk and intermediate products.

The results of this analysis suggest that noneconomic variables are quite important in explaining the processing level of U.S. agricultural exports. U.S. agribusinesses may have initially sought out markets that are familiar (areas where there are many U.S. tourists that are relatively urbanized), but they have also entered markets with different cultures. They may have been forced to face these differing social and cultural aspects of other countries to export high-valued food products. Western markets for high-valued food products have been saturated for a number of years, so the non-Western markets held the most profit potential. This will continue to be true in the future because many of the fastest growing future markets for U.S. agriculture will be in countries that are quite different culturally than the United States (Salvacruz and Reed, 19xx).

It is possible that what the U.S. agribusiness community needs is more help in internationalizing its marketing efforts — helping it become more familiar with different countries, their marketing systems, and their consumption patterns — rather than help through commodity programs or other price-based programs. High-valued U.S. exporters are likely concentrating on product differentiation, rather than lower costs of production.

If this education process doesn't expand into more agribusinesses, it may be more efficient for the United States to concentrate on production and exportation of bulk commodities, allowing foreign food manufacturers to produce the final, highly processed foods (or for U.S. firms to invest directly into food processing overseas, as they have been doing according to Reed and Marchant [1992]). This is exactly what the U.S. agricultural industry has concentrated on in the past — bulk commodity exports, though there are indications that this trend is changing.

Obviously there are many areas that need to be addressed further if the issues brought up in this chapter are to be fully understood. The analysis needs to incorporate foreign direct investment in overseas locations to investigate whether such investments complement or substitute for agricultural exports from certain processing levels. Further analysis of language differences might be fruitful. Finally, such an analysis might be useful with data on expenditures under the targeted export assistance (TEA) and market promotion program (MPP) to discover the extent that these U.S. government programs have increased agricultural exports by level of processing.

Notes

1. Economic variables are those measuring price or income effects, whereas noneconomic variables are those measuring social, cultural, or legal effects. There is no question, though, that these noneconomic variables have economic effects, as is explained later.

2. This breakdown is similar to one used by Traub in his study of export indices. However, world trade is broken down only into three categories, so when U.S. statistics are compared with world statistics, consumer-oriented trade data are used.

3. The developed countries were all European Community countries, Australia, Austria, Canada, Finland, Japan, Norway, New Zealand, Sweden, and Switzerland.

4. The two economic variables in the hybrid models were GDP and the appropriate tariff rate in mixture 1 and GDP and international reserves in mixture 2. The tariff rate was included because it is the only price factor in the modeling effort. The other two variables were chosen for the hybrid models because they consistently explained more variation in the dependent variables than the other economic variables.

References

Advertising Age. 62 (September 25, 1991).

Evans, C. "Expanding Export Markets for U.S. Agriculture Including Higher-Value Products." Unpublished Report Submitted to the President, 1990.

Foreign Agricultural Service. *Desk Reference Guide to U.S. Agricultural Trade.* USDA Agriculture Handbook, No. 683, March 1990.

General Accounting Office. *U.S. Department of Agriculture: Strategic Marketing Needed to Lead Agribusiness in International Trade.* General Accounting Office RCED-91-22, January 1991.

Jeannet, J., and H.D. Hennessey. *International Marketing Management: Strategies and Cases.* Boston: Houghton Mifflin Company, 1988.

Porter, M. *The Competitive Advantage of Nations.* New York: Free Press, 1990.

Reed, M., and M. Marchant. "The Global Competitiveness of the U.S. Food Processing Sector." *Northeastern Journal of Agricultural and Resource Economics* 22 (1992): 61-70.

Reynolds, B., S. Schmidt, and A. Malter. "Marketing High-Value Food Products in the Asian Pacific Rim." Agricultural Cooperative Service Research Report, No. 85, January 1990.

Salvacruz, J., and M. Reed. "Identifying the Best Market Prospects for U.S. Agricultural Exports." Forthcoming in *Agribusiness: An International Journal.*

Schulter, G., and W. Edmondson. *Exporting Processed Instead of Raw Agricultural Products.* USDA Economic Research Service Staff Report, No. AGES 89-58, November 1989.

Traub, L. *Value-Weighted Quantity Indices of Exports for High-Value Processed Agricultural Products.* USDA Economic Research Service Statistical Bulletin Number 827. August 1991.

United Nations. *Statistical Yearbook.* New York: UN, various years.

U.S. Central Intelligence Agency. *The World Fact Book.* Washington, D.C.: CIA, 1991.

U.S. Department of Commerce. *Statistical Abstract of the U.S.* Washington, D.C.: U.S. Department of Commerce, 1989.

World Bank. *World Development Report.* New York: Oxford University Press, 1990.

PART THREE

Conceptual Foundations and Assessments from Trade and Macroeconomic Theory

6

Conceptual Foundations from Trade, Multinational Firms, and Foreign Direct Investment Theory

Wilfred J. Ethier

Trade theory used to consist of several microeconomic ideas embodied in a few "standard" models—most notably the Heckscher-Ohlin-Samuelson model. Fifteen or twenty years ago, most research busied itself applying, modifying, or extending these models. But, increasingly, what were thought the principal implications of this theory were found dramatically contradicted (quite remarkably almost without exception) by actual world commerce. This tension between theory and reality has inspired most recent contributions to the theory of international trade.

This trade theory development is of fundamental importance for the theory of the multinational firm. There are two reasons. First, the "new" trade theory emphasizes just those features that seem central to multinationals: economies of scale, imperfect competition, strategic issues of various sorts, technological change, and diffusion. Second, and still much less appreciated, is the parallel fact that the theory of direct investment has its own dramatic tension between existing theory and apparent fact. This is fully as dramatic as that experienced by trade theory and promises to propel an analogous development.

This chapter starts by briefly reviewing the basic traditional ideas of trade theory, and the characteristics of trade that these ideas apparently lead us to expect are compared with the stylized facts. I then review and evaluate some responses to this and comment upon the implications of both the conventional theory and the new developments for the theory of the multinational firm. Next, actual features of direct investment are briefly described. The chapter concludes with descriptions of, and suggestions for, newer developments in the theory of the multinational firm.

Fundamental Ideas of Conventional Trade Theory

Comparative Advantage

Comparative advantage is the most fundamental idea of all. Suppose two economies (home and foreign) and two goods, food (F) and manufactures (M). Denote the *autarkic* price of F in terms of M by P for the home country and by P^* for the foreign. Suppose that $P > P^*$. Then, with relative prices equal to marginal rates of substitution, foreign residents could be induced to reduce their consumption of F by the receipt of less M than home residents would willingly pay for that F. And, with relative prices also equal to marginal rates of transformation, the foreign economy could produce more F by reducing M production by less than the home economy could increase it by cutting back F production. Thus gains can arise if the home economy exports manufactures in exchange for food.

This basis for trade is international *differences* in economic structure reflected in autarky price differences. Thus one might expect the gains from trade to be greatest between countries that are most *dissimilar*, and the temptation to trade also to be greatest for such countries. Since trade is here a matter of each country concentrating on producing what it 'does' relatively well — relative, both to what other countries do and to its own needs — trade might be expected to cause countries to specialize their production and to import goods quite different from what they produce and export.

The Heckscher-Ohlin Theory

The predominant explanation of comparative advantage is the factor endowments theory, expressed in the Heckscher-Ohlin-Samuelson model. Indeed until recently it was common to refer to this model as *the* "modern" theory of international trade. Suppose that food and manufactures can be produced by capital (K) and labor (L). The home and foreign economies differ only in their endowments of these factors,[1] which are immobile internationally. Call the home country the capital abundant one and good M the capital-intensive commodity. Then in autarky the wage-rental ratio (w/r) should be greater at home than abroad. This in turn implies that $P > P^*$: The capital abundant country has a comparative advantage in the capital-intensive good (the Heckscher-Ohlin theorem).

Trade will induce the home economy to shift resources from food to manufacturing. The contraction of the labor-intensive sector and expansion of the capital-intensive sector will, at initial factor prices, generate an excess demand for capital and excess supply of labor. Therefore w/r must fall to induce firms to substitute labor for capital and allow factor markets to clear. Indeed r must rise relative to both commodity prices and w must fall relative to both, provided the home economy is not driven to specialize completely in M production (the Stolper-Samuelson theorem). Otherwise the price of M would rise relative to both factor rewards, producing positive profits inconsistent with equilibrium, or the price of F would fall relative to both, producing losses inconsistent with continued operation of the sector. Thus international trade must raise the real reward of a country's relatively abundant factor and lower the real reward of the relatively scarce factor. Protection induces the opposite effect. Since international trade lowers w/r in the capital-abundant home economy and raises w^*/r^* abroad, relative factor prices in the two countries necessarily become more nearly equal as a result of trade. Indeed more can be said. As long as manufactures remain relatively capital-intensive, w/r is necessarily positively related to p, the relative price of F in terms of M, if both goods are produced in equilibrium. This relation is determined by technology and so applies to both countries. With p equalized across countries by trade, w/r is necessarily equalized as well, provided that both countries continue to produce both goods. This must be so if relative factor endowments are not "too" different. Furthermore, equalization of relative factor prices w/r necessarily implies equalization of absolute factor prices as well; otherwise one country would have uniformly higher costs than the other and thus not be able to compete in any sector, a condition inconsistent with equilibrium. In sum, free international trade causes factor prices to become more nearly equal internationally, and necessarily produces complete equality if relative factor endowments are sufficiently similar across countries (the factor price equalization theorem). If countries have identical factor prices, they will, of course, use the same techniques of production. The apparent implication is that, in a trading equilibrium, countries with roughly similar relative endowments will have similar factor prices and production structures while countries with more divergent relative endowments will not.

Suppose the home economy, producing both goods and exporting M, becomes more capital abundant. With commodity prices, and therefore factor prices, restrained by international markets, the additional capital can be utilized only if the capital-intensive sector, M, expands. But an expansion would also require more labor, which can come only from the F sector, which must therefore contract, freeing even more capital for the M sector. Thus, at constant commodity prices, an increase in the capital stock must cause a proportionally even larger increase in the production of the capital-intensive commodity and a lower production of the labor-intensive commodity (the Rybczynski theorem). This establishes a presumption that growth of a country's relatively abundant factor will generate an excess supply of exportables and an excess demand for importables, at the original terms of trade. That is, growth that makes a country *less* like the rest of the world should cause it to trade more and to experience a deterioration in its terms of trade; growth making the country more like the rest of the world should have the opposite effects.

Apparent Implications of Conventional Trade Theory

The following summarize the apparent implications of trade theory for the broad characteristics of world trade.

- Trade should be greatest between countries with the greatest differences in economic structure.
- The gains from trade should be greatest between countries with the greatest differences in economic structure.
- Trade should cause countries to specialize more in production and to export goods distinctly different from their imports.
- Countries should export those goods that make relatively intensive use of the countries' relatively abundant factors.
- Free trade should equalize factor prices between countries with fairly similar relative factor endowments but not between countries with markedly different endowments. The former countries should also employ relatively similar techniques and produce similar goods.
- Factor prices should be more nearly equal between countries with more liberal mutual trade.

These propositions are *not* — and the point can hardly be emphasized enough — inevitable consequences of the conventional theory of international trade. Each is true only subject to certain conditions, in some cases the conditions are severe, and conditions also exist under which some of the propositions are mutually inconsistent. But, nevertheless, the propositions, as a group, do accurately portray that view of the world economy that the conventional theory conditions one to accept as the norm.

Confrontation with Reality, I

For over thirty years the accumulation of knowledge — both quantitative and intuitive — about actual world trade has eroded the picture painted above. This process was effectively initiated by Leontief, who, in an attempt to muscle the actual data of U. S. trade and production into the literal 2x2x2 Heckscher-Ohlin-Samuelson straightjacket, found that American exports were significantly less capital-intensive than American import substitutes. Given the presumption that the United States was relatively capital abundant, this was in dramatic contrast to the Heckscher-Ohlin theorem.

Other key stylized facts emerged from investigation of the effects of economic integration in Western Europe. This integration was followed, for most countries, by an increase in both imports and exports across most sectors, rather than by an increase in specialization. Subsequently economists recognized this as part of a general trend not confined to the EEC. Furthermore, throughout the post-war period, trade among the developed countries — with relatively similar economic structures — increasingly dwarfed that between the developed countries as a group and the less developed countries. An overall picture emerged of world trade dominated by the intra-industry exchange of manufactures between roughly similar economies.

The factor–price equalization propositions, by contrast, do pretty well on balance (but this is not a claim of causality). Of the major theorems it is the factor-price equalization theorem that coexists most comfortably with reality. This is a fine irony that would have astounded writers in the fifties, when it was common to single out — sometimes with ridicule — this theorem as that one that was most abstract, most dependent upon unfulfilled assumptions, and least relevant to reality.

What emerges from this cursory review is a striking contrast between reality and the apparent implications of theory. This has come into focus only gradually in response both to accumulating empirical observation and to the evolution of commerce. The effect on trade theory has been profound. Until the early seventies theorists were primarily concerned with elaborating and tidying up the received body of theory. For the better part of the last decade, by contrast, the basic task has been to restructure the theory to imply a very different picture of world trade.

International Investment

Before turning to recent developments in trade theory, consider the ability of conventional theory to accommodate direct investment and the multinational firm. This is easy: The core of conventional theory simply has nothing to say. The neoclassical firm is a central agent in this theory. This agent's size and extent — including its possible extension across national borders — is indeterminate but also inconsequential.

The conventional theory imposes perfectly competitive markets and no externalities. In effect this rules out economies of scale and imperfect competition. As both seem to be prominent in the markets within which lurk multinational firms, it is probably a good thing that the traditional theory does not pretend to explain such firms. Both features are prominent in the "new" trade theory, so we shall turn to them presently. The theory described above also rules out international investment, which is likewise associated with direct investment (though not necessary for it). As this is easily introduced into the conventional theory, consider this now.

Introduce international investment as the transfer of a part of one country's endowment to its trading partner, in the Heckscher–Ohlin model described above. Then the following appear as immediate normal implications of the theory.

1. International investment should be stimulated by differences in factor endowments. This follows from the fact that differences in factor endow-

ments (and, from the factor price equalization theorem, *large* differences if there is free trade in goods) generate the differences in factor rewards that induce international factor movements.

2. International trade and international investment should be negatively correlated. This follows from the Rybczynski theorem: Since international factor mobility is effectively an example of mutual and opposite growth making two countries more alike, it should cause trade to contract. Commodity trade and factor movements should be substitutes: By driving factor prices together, commodity trade diminishes the temptation for factor mobility.

In fact, a large share of international investment takes place among the industrial countries, in defiance of both propositions. Substitutability of factor trade and goods trade is basic to the HOS model: It's almost tautological. But when trade is due to something other than different factor endowments, the two tend to be complementary.[2]

To see this, consider trade due to differences in technology. Suppose the HOS model with two completely identical economies, so there is no basis for either trade or factor mobility. Now suppose a small Hicks-neutral technical improvement in the home capital-intensive sector and a small technical deterioration in the home labor-intensive sector. Allow trade in goods but not in factors. The home economy will export the capital-intensive good and import the labor-intensive one. The home economy has a higher rent and lower wage than the foreign. If factor mobility is now allowed, capital moves into the home country and labor out. This strengthens the pattern of comparative advantage and so causes commodity trade to *increase* still more: Goods trade and factor trade are complements. The final equilibrium involves complete specialization by each country, with factor price equalization. Note also that a naive Leontief-type calculation would appear to support the Heckscher-Ohlin theorem. But this is misleading: The differences in factor abundance are not causing the commodity trade; rather *both* are caused by the difference in technology.

The Reaction

The pattern of world trade sits uncomfortably with comparative advantage. There is no *logical* inconsistency here, but this is of little comfort if comparative advantage addresses concerns only tangential to actual world commerce. But why should we care? Because comparative advantage has normative implications. These implications are important if actual trade is due significantly to comparative advantage, but may be misleading otherwise. For example, DC-LDC trade, between countries with significant differences in economic structure, would appear to promise large gains, and DC-DC trade would likewise seem to offer modest gains, if comparative advantage explains trade. This would seem to imply that we ought to try hard to integrate the

LDCs into the world economy and not worry over much about new marginal barriers to trade among the DCs. But this policy prescription could easily be disastrously wide of the mark if trade is in fact driven by considerations quite different from comparative advantage.

But what other candidates are there? There is in fact a very old one: increasing returns to scale. The fundamental idea behind comparative advantage is that countries trade in order to exploit their *differences*. Another possibility is that they might trade in order to *specialize*, that is, to become more productive by doing less but doing it better. Alternatively, they might trade to enlarge markets and thereby increase *competition*. Formal models help here.

Intra-industry Trade

Scale economies can offer a basis for intra-industry trade. Assume that increasing returns are external to the firm and internal to the national manufacturing industry. Also, disaggregate the manufacturing sector. That is, suppose that there are n distinct manufacturing subindustries, each producing a differentiated product. Unlike the Heckscher–Ohlin–Samuelson model, there is only one primary factor—labor. Each subindustry possesses the following technology. Each firm behaves as if it were subject to constant returns to scale. However, its productivity depends in fact on the size of the national subindustry. There are increasing returns to scale that are external to the firm and internal to the national industry. But the scale economies decrease with the size of the subindustry. The number n of manufacturing varieties will be determined endogenously.

Suppose for simplicity that all existing varieties of manufactures enter consumers' utility functions in symmetric ways. Given the symmetry in production, this implies that in equilibrium all varieties produced in a country are produced and consumed in equal amounts, say b. Suppose the varieties are imperfect substitutes and consumers prefer more variety to less: An increase in n and reduction in b leaving nb unaltered will improve welfare.

If the taste for diversity is sufficiently strong, and/or the returns to scale sufficiently weak, there should be an indefinitely large number of indefinitely small subindustries. Sufficiently strong scale economies, on the other hand, imply that n should equal unity. The most interesting possibility, therefore, arises in the intermediate case when there is an optimal size of each subindustry. Changes in the size of the manufacturing sector would change the number of manufacturing varieties and not the scale of each subindustry. Thus the industry as a whole would behave as if it had constant returns to scale.

Now allow international trade between two such countries, identical in every way except that foreign labor is more productive than home labor in food production. Then the home economy will export manufactures to the rest of the world for food. If the foreign economy produces only food, this inter-industry trade is all that will happen. But the foreign economy, though exporting food, might also produce some manufactures. This generates *intra-industry* trade: the varieties of manufactures produced at home should differ from those produced abroad, so each country exports a fraction of the output of each of its manufacturing subindustries. As the home economy has more subindustries, it runs an intra-industry trade surplus to exchange (*inter-industry* trade) for foreign food.

There is a sense in which intra-industry trade depends upon similarities across countries, while inter-industry trade depends upon differences. Suppose, for example, that both countries' labor are equally productive in the food industry, so that the two countries are completely alike. Then there s no basis at all for inter-industry trade, and each country will be self sufficient in food. But they should still trade: If they produce different assortments of manufacturing varieties they can both gain by exchanging these. All trade will then be intra-industry.

This picture can be generalized from the one–factor model. Consider the following separable production structure. Several factors determine a frontier of possible combinations of food production F and manufacturing resources m, and the latter are used, as described above, to produce manufactures. The factors might be two, with a Heckscher-Ohlin–Samuelson model essentially giving a production possibility frontier of F and m. In this case, relatively distinct factor endowments at home and abroad would produce mainly inter-industry trade, whereas endowment similarities would cause trade to be largely intra-industry. An international factor movement making the endowments more similar would shrink inter-industry trade and expand intra-industry trade. That is, the former is a substitute for international factor mobility whereas the latter is complementary to it. This *complementarity theorem* expresses the essence of recent theories of intra-industry trade.

The development of a theory of intra-industry trade like that summarized above has been one of the major accomplishments of the international trade theory of the preceding fifteen years. This theory was in fact developed in more complex models than that just discussed, no doubt because of the latter's tenuous relation to reality. But the most remarkable aspect of the development of this theory was that it was reached, almost simultaneously, by two independent lines of inquiry proceeding from two quite different starting points. I discuss each in turn.

International Economies of Scale

External economies have generally been identified with an increased division of labor allowed by a larger market: Adam Smith's pin factory and the Swiss watch industry for example. Less common examples involve more public information generated by a larger industry. In principle none of these requires an industry to be physically located in one place. A dispersed industry can realize a great division of labor if intermediate components can be shipped from place to place; public information can be dispersed within the industry if communication is efficient. If so, what matters is the global size of the industry, not its geographical concentration.

This suggests that the returns to scale depend upon the size of the *world* industry, not the national industry. This is what is meant by *international* returns to scale.[3] To uncover the implications, suppose, as above, that capital and labor are combined, in a Heckscher-Ohlin framework, to produce food, F, and m. m is an index of the scale of operations of the national manufacturing industry, subject to increasing returns to scale. With *national* returns to scale, national manufacturing production M was related to m by

$$M = k \cdot m \quad \text{where } k = k(m), k' > 0 .$$

With *international* returns to scale, on the other hand, we have instead

$$M + M^* = k \cdot (m + m^*) \quad \text{where } k = k(m + m^*), k' > 0$$

Here an asterisk refers to the foreign country.

At first glance it might seem that this complicates matters enormously. National production possibility frontiers between final goods are not even defined, because productivity in each country's manufacturing industry depends upon the size of the other country's manufacturing industry. But the situation becomes almost transparent with a focus on patterns of resource allocation rather than on goods.

Consider the world production possibility frontier between food and manufactures. A point on it can be found by maximizing world manufacturing production for a given feasible volume of world food production, that is, by

$$\text{maximizing: } M + M^* = k(m + m^*) \cdot [m + m^*]$$

$$\text{subject to: } T(m) + T^*(m^*) = \text{some specified value.}$$

Here T and T^* denote the home and foreign production possibility frontiers between food and manufacturing resources. It is immediately clear that $M + M^*$ will be maximized by maximizing $m + m^*$. That is, efficient patterns of world activity in food and manufactures correspond to efficient patterns in food and manufacturing resources, ignoring the scale economies.

International economies of scale imply a theory of the intra-industry exchange of intermediate goods between relatively similar economies. A dispersed industry can realize the benefits of a large division of labor if intermediate goods can be shipped within the industry. Thus the more nearly equal in size m and m^* are, the greater the volume of intra-industry trade in manufacturing components.

The picture of intra-industry trade, and of its relation to inter-industry trade, is very much like that of the previous subsection. With identical homothetic demands across countries, the pattern of inter-industry trade and specialization is determined in the familiar Heckscher-Ohlin fashion. If food is relatively labor-intensive, the relatively labor abundant country will export food for manufactures. This inter-industry trade will comprise all trade if the endowment disparity causes the labor-abundant country to specialize completely in food. Endowment differences reduce the incentive for inter-industry trade but cause the integrated manufacturing industry to be divided relatively evenly between countries, thereby inducing intra-industry trade. It should be clear that the Complementarity Theorem holds.[4]

Product Differentiation

Several general equilibrium theories of monopolistic competition were developed during the seventies. It was natural to apply these theories to intra-industry trade, and that application constituted the second independent approach.[5]

Suppose again that capital and labor combine to produce food (F) and m, with the latter transformed into manufactures. There are now no scale economies for the manufacturing industry overall (so $k = k^* = 1$), but that industry is disaggregated into an endogenously determined number n of subindustries, producing different varieties of finished manufactures. Each subindustry operates under increasing returns to scale, but the scale economies are now internalized by individual firms. The manufacturing industry is modeled as one of *monopolistic competition*, with each variety produced by a single firm operating under increasing returns to scale. The firm is a mini-monopolist, maximizing its profit by equating marginal cost to marginal revenue, given the prices charged by its many fellow manufacturing producers. Profits are driven to zero by the entry and exit of new firms (with new varieties), that is, by adjustment of n.

It should be apparent, after a little thought, that the model of this section, with either assumption about preferences, generates the same picture of intra-industry trade that emerged in the previous subsection. That is, comparative advantage determines the inter-industry exchange of food for manufactures, similarities between countries induce the intra-industry exchange of manufacturing varieties, and the Complementarity Theorem holds. Even more striking, in view of the radical difference in starting points, is the fact that international returns to scale and product differentiation generate essentially the same theory of intra-industry trade even though one applies to producer goods and one to consumer goods. A formal model of either can be reinterpreted as a formal model of the other [see Ethier (1982)]. It's worth pointing out, however, that if the preference structures described in this subsection are reinterpreted as descriptions of a technology transforming a collection of differentiated intermediate goods into a single final output, horizontal product differentiation seems more "realistic" than individualized differentiation, just the reverse of the common belief. And trade in producer goods is more prominent in fact than trade in consumer goods.

Oligopoly as a Basis for Trade

Consider the same production model, except that the number n of manufacturing producers is small and fixed, and all varieties of manufactures are regarded by consumers as perfect substitutes. There may be national increasing returns in manufacturing production, but they are internalized by firms.

Suppose that the home and foreign economies are identical, and that initially they do not trade and have identical autarky equilibria. Then there is no comparative-advantage basis for trade, and no product differentiation or international returns to scale. Nevertheless, the removal of trade barriers will not be without effects.[6] Each manufacturing producer will experience an increase in the number of its competitors from n-1 to $2n$-1. Exactly what effects this will have will depend, as is so often the case with oligopoly theory, on the firms' strategy variables and on the equilibrium concept.

Suppose equilibrium is Cournot-Nash and that each firm decides how many manufactures to supply to the single world manufacturing market: Under fairly reasonable conditions, the price of manufactures will be lower and the supply greater without trade barriers than with them, because of the greater competition faced by each firm. Thus oligopoly provides an independent basis for trade. Indeed no actual trade need take place! In the symmetric equilibrium of the present model, for example, both countries remain self-sufficient in both goods. The two countries in effect trade competition, not goods.

Suppose now that the strategy variable of each firm is the number of manufactures to sell in *each* country. (Suppose also that there is no secondhand market in manufactures so that firms sell directly only to consumers). Equilibrium will clearly be as before, except that each firm will sell half its output in each market. There will thus be two-way trade in identical products, that is, intra-industry trade that cannot be made statistically to disappear by defining an industry narrowly enough.

Next suppose a modest cost of shipping manufactures between national markets. Foreign firms will now be at a disadvantage, relative to domestic ones, in each market since their goods bear the transport cost. Under reasonable assumptions the equilibrium will be one in which each firm sells part, but less than half, of its output in the other country. Each firm will charge a lower export price (by the amount of the per unit transportation cost) than domestic price.[7]

Thus there will still be two-way trade, but now it is costly since resources are used shipping manufactures both ways, the price for the better allocative efficiency that comes from the increased competition. Of course it would be even better to have the competition without the transportation. This is what happens if the strategy variable is total supply rather than supply to each market.

Oligopoly Profits

The manufacturing producers need not earn zero profits in equilibrium. This distinguishes oligopoly from the market structures we dealt with earlier. These profits are part of national income and therefore must be considered in an analysis of the welfare implications of trade or of trade policy.[8]

To isolate the profit aspect, suppose that the home and foreign countries do not trade at all with each other but rather with a third country that itself produces only food, and that only food is consumed in the home and foreign countries. This removes the need to consider the effect of a distorted manufacturing market on home and foreign consumers.

Suppose initially that $n = n^* = 1$ so that the home and foreign firms constitute a duopoly in the third country. Suppose they establish a Cournot-Nash equilibrium, N. This will not yield the home firm as much profit as a Stackelberg equilibrium S with the home firm as leader and the foreign as follower. But the home firm cannot achieve S if the two firms are treated symmetrically: An announcement by the home firm that it would produce its Stackelberg quantity no matter what is not credible. The only pair of credible announcements are of course those corresponding to N.

The home government enters the analysis at this point. The government can tax (or subsidize) manufacturing exports — equivalent to a production tax, since all manufactures are exported. The government wishes to maximize national income, equivalent to maximizing the profits of the home manufacturing firm. Of course this is just what that firm wants to do, so there is a role for government action only if the government has some power the firm does not.

The power that has been discussed is that of credible commitment. The government can announce a tax on exports and both firms believe that it will in fact be implemented. A per unit export tax in effect shifts the home firm's marginal cost curve up by that amount. The marginal revenue curve corresponding to any given foreign output is not affected, so the home output that maximizes home profit given foreign output will fall. That is, an export tax shifts the firm's isoprofit curves to the left, and an export subsidy shifts them to the right. The government should *subsidize* exports just enough so that S becomes the Cournot-Nash equilibrium.

There are really three important ideas here, sensitive in varying degree to the special assumptions. The first is simply that oligopoly profit constitutes a new element of national income and thus a new concern of trade policy. This point is clearly quite general.

The second is that, with oligopoly, a new potential for trade policy comes from the fact that the government can make credible promises that individuals cannot. This is more problematic. An assumption that the government can credibly commit itself to a tax rate but that a firm cannot credibly commit itself to any export level may be arbitrary, but the general notion that the domain of credibility of the government might be different from that of private agents is a valuable insight.

The final important idea is that oligopoly might render it in the national interest to subsidize trade rather than restrict it. This conclusion, however, is quite fragile.[9] For example, it depends on the assumption of just one home manufacturing firm. In this case we want the home firm to become more competitive with its foreign rival and expand its share of the market and so its share of profits. But if there are two or more home firms, they compete against each other as well as against the foreign firm. We would like to *restrict* this competition to improve the terms of trade. If n becomes large enough, the manufacturing industry will become competitive and the familiar optimum tariff argument will imply an export tax. Thus we would expect that, as n rises, the optimal subsidy falls and then becomes a tax.

Another reason is that the conclusion is sensitive to the equilibrium concept. For example, with Bertrand–Nash equilibrium instead of Cournot-Nash, the optimal policy is likely to be an export *tax*. Evidently the conclusion that trade should be subsidized rather than taxed is very sensitive to the equilibrium concept.

Direct Investment

Although traditional trade theory has said little about the multinational firm, a huge, less formal, theoretical literature has existed for a long time. This literature is concerned in part with why multinational firms should exist at all in the face of presumed costs of operating across national frontiers.[10] The theory has three components. First, the firm should possess an *ownership* advantage, such as a patent or some managerial or organizational ability, to exploit in several national markets. Second, *locational* considerations should ensure that the firm does not find it attractive to concentrate all operations in one country and export to others. Finally, the *internalization* of international transactions must be preferable to the use of markets. The firm, for example, should find it advantageous to conduct foreign manufacturing itself rather than to license a foreign firm to do it.

Multinational Oligopoly

Consider the following model.[11] Two identical countries produce food and manufactures, that each exports manufactures to a third country for food, that there are but two manufacturing firms (one in each exporting country), and that the third country produces only food. But now elaborate as follows. Food is produced by labor and land, whereas manufactures are produced from labor and capital. The capital stock in each manufacturing country is owned by its manufacturing firm, and the third country has no capital. The third country is relatively labor abundant in the sense that, in the trading equilibrium, its wage is lower than wages in the home and foreign countries.

If trade takes place, and no international factor mobility is allowed, the earlier discussion applies completely. If capital were to become internationally mobile, each manufacturer would move at least some of its capital stock to the third country and produce manufactures there, displacing exports. If costs are equalized before the entire capital stock has been moved, the two manufacturers will each produce in two countries: They will have become multinational firms.

This is, of course, just the standard factor-endowments story of factor movements substituting for commodity trade.[12] Now take a different tack. Distinguish real factor movements from direct investment by assuming that all factors, including capital, are internationally immobile, but that manufacturing firms can operate in either country. Interpret capital as the input to an intermediate activity, such as research or management, that can be conducted at a distance from the productive units that utilize it. That is, assume that each firm can produce manufactures by combining its capital with labor located in any country. Some examples of such activities, such as research, have at least a partial public good aspect. It's natural to try to capture this by supposing that manufacturing production entails international economies of scale that are internal to the firm.

With wages lowest in the third country, each manufacturer will now establish production there and become multinational. Suppose that third-country wages are never driven up to home and foreign levels. Then each manufacturing firm is a multinational employing capital in the source country and labor in the host country. Trade in manufactures ceases, with the home and foreign countries importing food as payment for their repatriated profits. Clearly direct investment has substituted for trade in the same way that real capital movements did in the previous example.

Note also that the earlier discussion of trade policy holds here exactly. That is, an increase in the profits of the home-based multinational will raise home national income. If the government can precommit itself while the firm cannot, there is scope for a direct investment policy, and if equilibrium is Cournot-Nash the optimal policy will call for a subsidy on direct investment abroad.

Now suppose that the home and foreign economies consume both goods and deal with each other. Equilibrium is Cournot-Nash and each country determines strategy variables for each market, rather than for the world as a whole. Then each firm will supply half of each market, as above. There will be two-way trade in identical products and no reason for direct investment.

Next, again add a transportation cost. Now there is an incentive to go multinational. Each firm will continue to supply one-half of each market but by local production. This cross-penetration is not pointless: It allows the benefits of increased competition without the costs of shipping manufactures back and forth. This is another case of direct investment substituting for trade. Two-way direct investment arose simply because the initial trade model featured two-way trade.

Multinational Monopolistic Competition

Next turn to monopolistic competition.[13] Suppose the manufacturing sector in each country is again composed of an endogenous number (n and n^* respectively) of firms producing differentiated products. Each firm employs capital at home together with labor, in either country, to produce its unique variety, subject to increasing returns to scale. Consumers have a preference for variety that allows the existence of many manufacturing firms in equilibrium.

Suppose that the two countries are identical and allow trade and direct investment. With identical endowments and homothetic preferences there is no basis for the inter-industry exchange of food and manufactures, and factor price equalization ensures that there is no motive for direct investment. International exchange will consist of the intra-industry trade of manufacturing varieties.

Now let endowments differ slightly, with the home country relatively capital abundant. Then it will export the capital-intensive commodity, manufactures, for food; with world manufacturing production somewhat more concentrated at home, there are fewer varieties to import. Thus inter-industry trade displaces intra-industry trade, as before. As modest endowment differences still cause trade to produce factor price equalization, direct investment still does not take place.

Larger endowment differences would prevent trade from itself establishing factor price equalization. Foreign wages would tend to fall below home wages, and at least one country would tend to specialize. The lower foreign wage would induce home manufacturing producers to hire foreign labor and thereby become multinational firms. Foreign manufacturing producers remain national, if some foreign capital remains in the manufacturing sector. If home firms employ labor in both countries wages must be equal; the direct investment will have prevented factor price disparities from emerging. But if the endowment difference is sufficiently pronounced, home manufacturers will employ only capital at home and factor prices will diverge.

Thus endowment differences generate direct investment, which cannot occur when trade alone would equalize factor prices. The formation of multinationals is associated, in this model, with a decline in intra-industry trade relative to inter-industry trade. Two-way direct investment can never take place, but the introduction of barriers to intra-industry could induce it, just as in the preceding section. In this model, direct investment is in effect a proxy for real international capital mobility.

Confrontation with Reality, II

Patterns of direct investment bear a striking analogy to patterns of trade discussed earlier. The largest part of direct investment, and the fastest-growing part, is between developed market economies, rather than from the DCs to the LDCs. This is shown in Table 6.1.

Furthermore, this investment is increasingly *two-way* and also increasingly *intra-industry*. Also as with trade, there has been recent dramatic growth in direct investment originating from the newly industrializing countries. Finally, since World War II there has been a dramatic change in focus from raw material based sectors to manufacturing.

The next two sections discuss, in increasingly preliminary fashion, areas in which existing theories of direct investment have not yet made much progress but which the facts of world commerce suggest are central: Internalization and the international dissemination of technological advance. Each topic is the subject of a large literature but not in the context of the multinational firm. Thus our treatment will necessarily be suggestive and incomplete.

Internalization

So far direct investment has functioned the way real capital movements would. This has given a rich variety of behavior, but these multi-plant scale economy models of the multinational firm are unsatisfying in two ways.

First, they simply do not address the central phenomenon of the larger part of direct investment taking place between similar economies, much of it two-way. The second point is conceptual. The dominant paradigm of direct investment views it as determined by a coincidence of ownership, locational, and internalization considerations. The above models concerned in detail the

TABLE 6.1 Patterns of International Direct Investment, 1981-85 (average annual flows in billions of $)

Investment into:	Investment from:		
	DCs	LDCs	Total
DCs	$35 (74%)	1 (2)	36 (76)
LDCs	12 (24)	0 (0)	12 (24)
Total	47 (98)	1 (2)	48 (100)

Source: UN Centre on Transnational Corporations

first two of these but took it for granted that internalization was advantageous. This is arbitrary. Ownership and locational considerations have always been basic to trade and imperfect competition without direct investment; the internalization issue is the new element introduced by consideration of the multinational firm and where the conceptual value added of studying direct investment lies.

So consider internalization.[14] Whether a transaction should be internalized is basically a matter of the exchange of information between agents, and this becomes substantive only when some agents face uncertainty. Start with the product differentiation model just discussed. Suppose, though, that finished manufactures are nontraded in the sense that the labor employed to produce them must reside in the country where they are consumed. This means that each variety of manufacturing will involve labor in both countries. This assumes away locational considerations and allows a focus on the internalization question of whether the labor used in the two countries will be employed by a single firm or by two firms. To introduce uncertainty, suppose that each variety involves an upstream activity, which uses only capital and so is located in the source country, and a downstream activity, which uses only labor and so must locate in both countries. The upstream activity involves research and, after the results of the research are known, producing the sensitive innards of the manufactures. These parts are then shipped to the downstream firms, which use labor to produce finished manufactures for consumers. The cost of the innards produced upstream might be either high or low and is unknown until the outcome of the research project. The more research that is done, the greater the chance that the cost will turn out to be low. The upstream firm must commit itself to research, and the downstream firms to labor, before the outcome of the research project is known. The choice of whether to produce innards can be made ex post.

The basic problem is to distinguish those cases in which the upstream and downstream units can, without cost, deal with each other at arm's length from those in which they cannot, and therefore must internalize their transactions and become a single firm. Presumably there is some ex ante contract, calling for the upstream firm to conduct a certain amount of research and to produce a certain quantity of innards for each possible outcome of the research project, and calling for the downstream firms to employ certain quantities of labor and to make certain payments to the upstream firm, that will maximize joint profit.

There are various possible assumptions about when such an optimal contract can be implemented at arm's length. For example, the upstream firm might be unwilling to let independent downstream firms observe its research effort or the outcome of that effort: This is, after all, the source of its ownership advantage. This could then give rise to an incentive compatibility problem, and we could require that any arm's length contract be incentive compatible whereas an integrated firm would not be subject to such a constraint. This constraint would not in fact be binding if the independent firms were risk neutral and if it were feasible to write contracts covering all conceivable states. Otherwise there would be circumstances under which the constraint would bind and when we would therefore expect multinational firms to emerge.

Another possibility would be to require a contract to be "simple" to be implementable. For example, in industries where research is important or where tastes and techniques are subject to significant change, the number of conceivable states of nature might be so vast that any feasible contract would necessarily be incomplete or insensitive to some contingencies. It's relatively simple to pursue this in the present framework, so do so. Since there are only two possible states, call a contract that calls for behavior to differ across states "complex" and one that does not "simple." That is, the upstream and downstream firms are constrained to implement a state-invariant contract if they remain at arm's length but can free themselves of the constraint by integrating.

When will the constraint be binding, that is, when will the optimal contract vary across states? Production of each variety entails fixed costs (research and employment) that must be met whether any innards are actually produced or not, and variable cost (innard production) that can be avoided by not producing. Manufactures are capital-intensive, so we would expect, other things equal, that if rents are sufficiently low the optimal contract will call for innards to be produced in both possible outcomes. (If rents are low enough, variable costs can be covered in either case.) But if rents are high, it is likely that the contract would call for production to occur with a favorable outcome, producing a positive profit, but not with an unfavorable outcome. Thus we would expect the optimal contract to be simple when rents, in the host country, are low and to be complex when they are high.

But manufactures are capital-intensive, and the relatively capital abundant country will be operating (upstream) manufacturing firms. Thus we would expect rents in the host country to be relatively high when endowment differences are modest and relatively low when they are great. That is, direct investment will be induced by *similarities* in endowments rather than by differences, as in the previous subsections.[15] Sufficient endowment similarity could induce two-way direct investment.

Internalization, when modeled in this general way, produces results quite different from those of multi-plant economies models of direct investment that abstract from internalization. They also come to grips with the phenomenon of two-way direct investment between similar economies. It remains to be seen, however, what happens when internalization is modeled differently.

Technology Diffusion and Internalization

This section addresses internalization in relation to technological competition and dissemination.[16] Multinational enterprises are closely linked to knowledge-based capital, whose services are easily transported between distant locations (managers and engineers visiting plants) compared to physical capital, and which often facilitates multi-plant production (blueprints or chemical formulae are costlessly supplied to additional plants).

The Basic Model

Assume two countries, H and R, each endowed with a single factor of production, labor, immobile between countries. Firms in H use labor to conduct research and to produce goods; firms in R only produce goods.

Firms in H enter a two-period race to develop a new product; the winner captures the exclusive knowledge of how to produce the new product plus a plant in H to do so at constant marginal cost. The product remains new for two periods until the appearance of the next new product. Assume that the firm must supply the H market from its plant located there and that it can prevent anyone else from producing the product in H for the two-period duration of its newness. The firm can supply the R market either by exporting from its plant in H or by local production in R. The latter might be done by the H firm itself employing labor in R at a subsidiary, or by a firm in R licensed by the home firm.

Knowledge of how to produce a new product disseminates gradually. In the first period, only the H firm that developed the product knows how. Anyone involved in producing the good in the first period can produce it in the second. Thus first-period franchisees or subsidiary employees can now produce it themselves. After the second period, knowledge becomes common: Any firm in either H or R can produce it, and it ceases to be new. Thus the MNE may need to choose between maximizing immediate profit and the possible dissipation of its proprietary asset.

As long as the product remains new, production in either location involves a per-period fixed labor cost G plus a constant marginal cost of one unit of labor. Since the firm must supply H from its home plant, the fixed cost there is not relevant to exports, but exporting does involve an additional constant transfer cost t per unit, in terms of H labor.

The services of the knowledge-based capital resulting from the firm's research project can be costlessly supplied to foreign producers, so G reflects additional input needed for production of a good that, because of its newness, may require unusual facilities or monitoring independent of the length of the production run. These inputs will no longer be required when production becomes standardized, that is, when the good becomes old. Higher values of G indicate that the public good aspect of knowledge is less important (i.e., knowledge is a less pure public good).

Let the monopoly rent that could be earned by production in R be denoted R. If R is supplied by production in H, the marginal cost is instead $1 + t$ but the fixed cost G need not be incurred; let this monopoly rent be denoted E. Let R^* denote total duopoly profit of two identical firms, both located in R, so that each earns exactly half of R^*. Finally, denote that Cournot-Nash duopoly profit for two firms together as E^*, if the only difference between the two is that one produces in H and one produces in R. Let a denote H's share of E^*. It can be shown that $R \geq (1-a)E$, that R exceeds R^*, and that $(1-a)E^* \geq R^*/2$ (5). On the other hand, the model implies no necessary relation between E and R^*.

Choice of Supply Mode

Consider the H firm's choice of how to supply the R market while its products remain new. Two crucial assumptions are maintained throughout. (1) Only a single H firm can initially produce new goods, but many R firms compete with each other, so H can dictate the terms of any agreement. (2) H cannot prevent a first-period partner from producing new goods in competition with H in the second period. Nor can the R partner prevent H from exporting on its own or from taking on a new partner in the second period.

H might simply export during both periods, earning E in each for total discounted earnings of $E(1+d)$. Or H might export in the first period and license a firm in R during the second. The licensee during the second period could earn R, which H could extract as a license fee, since there are many potential partners in R. Thus H earns $E + dR$ over the two periods.

Alternatively, H might license a firm in R for the first period, during which H earns the fee Q, and then compete with its former licensee, earning the fee Q^* by licensing another foreign firm during the second period.

The final possibility is for H to become an MNE by establishing a foreign subsidiary. H would agree to pay its employees in R total compensation of C_1 and C_2 in the respective periods. Let the two-period payments received by the firm from the subsidiary be given by Q_1 and Q_2, with a present value of $Q_1 + dQ_2$. The subsidiary's employees then receive $C_1 + dC_2 = (R - Q_1) + d(R - Q_2)$. The MNE would be permanent arrangement, providing for the exploitation by its employees in R of any future new goods developed by H. Depending on circumstances, H might be able to capture the entire surplus $R(1+d)$ or it might not. Describe the former case as a *rent-capturing*, multinational enterprise (RC MNE) and the latter as a *rent-sharing* one (RS MNE).

Table 6.2 lists the possible alternatives with their implied payoffs.[17] The earnings entries record total two-period discounted earnings of the respective firms. R_1 refers to foreign participants in an arrangement during period 1, and perhaps in period 2 also; R_2 refers to a foreign participant only in the second period.

For any arrangement to be feasible it is necessary that it be in each party's interest at each stage. For example, if employees in F would benefit by deserting an MNE in the second period to produce the good themselves, this will be foreseen and the MNE will not be established in the first place. That is, assume no exogenous enforcement agency. Among the feasible outcomes, the H firm

TABLE 6.2 Alternative Arrangements

Arrangement	H Earnings	R_1 Earnings	R_2 Earnings
Exporting (X)	$E(1+d)$	0	0
Exporting, then Licensing (XL)	$E + dR$	0	0
Successive Licenses (LL)	$Q + dQ^*$	$R-Q + d(R^*/2)$	$d((R^*/2) - Q^*$
RC MNE	$R_1 + dR_2$	0	0
RS MNE	$R-C_1 + d(R-C_2)$	$C_1 + dC_2$	0

implements the one that maximizes its own profit. Different parameter values support different outcomes. For example, if $R < E$ there exists no contract that some R would accept that can give H a payoff greater than or equal to $E(1+d)$. Thus H will export in both periods. More generally, circumstances under which each arrangement is feasible can be deduced.

The Influence for Basic Parameters

The above description allows an analysis of how the values of the basic parameters determine the choice of supply mode. Table 6.3 summarizes the influence of locational considerations in this model. The parameters of interest are G, t, and w. High values of G and w, and low values of t, should be thought of as analogous to *dissimilarities* between countries as discussed earlier: They stimulate trade.

The influence of locational factors is more complex now than in the preceding section, a reflection of the fuller interplay among ownership (the fruits of research and the size of G), locational (t and w), and internalization (prevention of dissipation of the ownership advantage) considerations. Traditional locational situations are necessary for production in R: t must be in high enough and/or w low enough. But once this is so, internalization aspects require that t not be too high and w not be too low to allow the credible threats necessary to sustain a relationship between partners in H and in R.

TABLE 6.3 Circumstances Leading to Alternative Equilibria (H, M, or L indicates whether high, medium, or low parameter values make the respective arrangements the most likely outcome)

	G	t	w
X	H	L	H
XL	H, M	?	H
LL	L	H	L
RS MNE	M, L	M	L, M
RC MNE	H, M	L	M

Concluding Remarks

This paper has attempted to convey something of the flavor of current research into the theory of international trade and direct investment. I have tried to be suggestive rather than comprehensive. The discussion has necessarily been at time incomplete. Also many current topics of importance have necessarily been ignored completely. These are too numerous even to mention.

But two examples can be cited, even if only to tantalize with what we have missed. One is the vast topic of sovereign default and expropriation. This is clearly related to our discussion of technology transfer. The other is the complex of issues involving national tax systems in the presence of direct investment. These are central to an appreciation of the welfare consequences of the multinational firm. They also bring in a host of subsidiary issues, such as transfer pricing.[18]

Notes

1. In particular, production functions are identical across countries and characterized by constant returns to scale, and tastes are identical across countries and homothetic.

2. The detailed argument may be found in Markusen (1983).

3. See Ethier (1979, 1982) and Helpman (1984).

4. Indeed this is the context in which the Complementarity Theorem was originally developed. See Ethier (1979, p. 19).

5. The early contributions — independent of each other as well as of the international scale economies development — were Dixit and Norman (1980), Krugman (1979) and Lancaster (1980).

6. See Brander and Krugman (1983) and Helpman and Krugman (1985, ch. 5).

7. Brander and Krugman (1983) refer to such a situation as "reciprocal dumping."

8. This was the topic of a series of papers by Brander and Spencer (1984, 1985).

9. This is the basic theme of Eaton and Grossman (1986).

10. This approach derives from Coase (1937). See, for example: Hymer (1960), Caves (1971, 1982), Buckley and Casson (1976) and Dunning (1981).

11. For the multi-plant economies model of the multinational firm, see Markusen (1984).

12. See Caves (1971) and Jones, Neary, and Ruane (1983).

13. Product differentiation was introduced into the multi-plant economies model of direct investment by Helpman (1984). See also Helpman (1985) and Krugman and Helpman (1985, ch. 12, 13).

14. See Ethier (1986). Williamson (1975) contains a general discussion of relevant transactions cost issues.

15. See Ethier (1986) for details.

16. The following is based on Ethier and Markusen (1991).

17. Under the assumptions of this model, H will never choose to license a firm in R in period one and then to export in period two.

18. For discussions of issues relating to transfer pricing, see the papers in Rugman and Eden (1985).

References

Brander, J., and P. Krugman. "A 'Reciprocal Dumping' Model of International Trade." *Journal of International Economics* 15 (1983): 313-21. (Reprinted *Imperfect Competition and International Trade*, edited by G. Grossman. Cambridge, MA: The MIT Press, 1992).

Brander, J.A., and B.J. Spencer. "Tariff Protection and Imperfect Competition." In *Monopolistic Competition in International Trade*, edited by H. Kierzkowski, 194-206. Oxford: Oxford University Press, 1984. (Reprinted in *Imperfect Competition and International Trade*, edited by G. Grossman. Cambridge, MA: The MIT Press, 1992).

————. "Export Subsidies and International Market Share Rivalry." *Journal of International Economics* 18 (1985): 83-100.

Buckley, P.J., and M. Casson. *The Future of the Multinational Enterprise.* London: Macmillan, 1976.

Caves, R. E. "International Corporations: The Industrial Economics of Foreign Investment." *Economica* 38 (1971): 1-27.

————. *Multinational Enterprise and Economic Analysis.* Cambridge: Cambridge University Press, 1982.

Coase, R. H. " The Theory of the Firm." *Economica* 4 (1937): 386-405.

Dixit, A. K., and V. Norman. *Theory of International Trade.* Cambridge, MA: Nisbet and Cambridge University Press, 1980.

Dunning, J. H. "Explaining the International Direct Investment Position of Countries: Towards a Dynamic or Developmental Approach." *Weltwirtschaftliches Archiv* 117 (1981): 30-64.

Eaton, J., and G. M. Grossman. "Optional Trade and Industrial Policy under Oligopoly." *Quarterly Journal of Economics* 101 (1986): 383-406. (Reprinted in *Imperfect Competition and International Trade*, edited by G. Grossman. Cambridge, MA: The MIT Press, 1992.

Ethier, W. J. "Internationally Decreasing Costs and World Trade." *Journal of International Economics* 9(1979): 1-24.

————. "National and International Returns to Scale in Modern Theory of International Trade." *American Economic Review* 72(1982): 389-405. Reprinted in *Imperfect Competition and International Trade*, edited by G. Grossman. Cambridge, MA: The MIT Press, 1992.

————. "The Multinational Firm." *Quarterly Journal of Economics* 10 (1986): 805-33. Reprinted in *Imperfect Competition and International Trade*, edited by G. Grossman. Cambridge, MA: The MIT Press, 1992.

Ethier, W.J., and J.R. Markusen. "Multinational Firms, Technology Diffusion and Trade." International Economics Research Center Discussion Paper, University of Pennsylvania, 1991.

Helpman, E. "A Simple Theory of International Trade with Multinational Corporations." *Journal of Political Economy* 92(1984): 451-71.

————. "Multinational Corporations and Trade Structure." *Review of Economic Studies* 52(1985): 443-57. Reprinted in *Imperfect Competition and International Trade*, edited by G. Grossman. Cambridge, MA: The MIT Press, 1992.

Helpman, E., and P. R. Krugman. *Market Structure and Foreign Trade.* Cambridge, MA: The MIT Press, 1985.

Hymer, S. *The International Operations of National Firms.* Ph.D. diss., Massachusetts Institute of Technology, 1960.

Jones, R. W., J. P. Neary, and F. P. Ruane. "Two-Way Capital Flows: Cross Hauling in a Model of Foreign Investment." *Journal of International Economics* 14(1983): 357-66.

Krugman, P. "Increasing Returns, Monopolistic Competition, and International Trade." *Journal of International Economics* 9(1979): 469-80.

Lancaster, K. "Intra-Industry Trade under Perfect Monopolistic Competition." *Journal of International Economics* 10(1980): 151-76.

Markusen, J. R. "Factor Movements and Commodity Trade as Complements." *Journal of International Economics* 14 (1983): 341-56.

———. "Multinationals, Multi-Plant Economies, and the Gains from Trade." *Journal of International Economics* 16(1984): 205-26.

Rugman, A. M., and L. Eden, eds. *Multinationals and Transfer Pricing.* New York: St. Martin's Press, 1985.

Williamson, O. *Markets and Hierarchies.* New York: Norton, 1975.

7

International Competitiveness: Implications of New International Economics

Steve McCorriston and Ian Sheldon

Introduction

In recent years, concern has been expressed, in both popular and academic circles, about the competitiveness of the U.S. economy. The source for much of this concern has been, to a considerable extent, associated with the U.S. trade deficit and, related to this, the bilateral trade deficit the United States has sustained with Japan. Consequently, there has been much discussion and criticism of other countries' trade and industrial policies (particularly those of Japan and the European Community) on the grounds that U.S. exporters have had considerable difficulty increasing or maintaining market share abroad and that other countries' policies have given competitors to the United States assistance in penetrating the U.S. market. This, in turn, has given rise to demands for more U.S. government intervention, particularly in what is seen as a key sector of the economy, the high technology sector. Explicitly, there have been demands for "managed trade" (see, for example, Dornbusch, 1990 and Tyson, 1990) that would involve establishing "rules of the game" for trade in certain sectors. The ultimate aim of "managed trade" would be to promote U.S. access to overseas markets (particularly, but not only, Japan) in return for continued, but controlled access to the U.S. market. Others have called for internal measures to increase the United States' ability to compete in high technology industries (see Jarboe, 1985 and Tyson, *op.cit.*). Such domestic intervention may involve, for example, subsidies to research and development (R&D) and other instruments of industrial policy.

This chapter aims to provide a perspective on the desirability of such policy options as an aid to promoting a country's competitiveness. Specifically, we will discuss the insights that recent developments in trade theory offer in understanding the links between policy and competitiveness. There are, however, two points that should be noted as a preamble to the discussion. First, since competitiveness is not solely a concern of the United States, we will attempt to keep the discussion as general as possible. One only has to recall the not dissimilar concerns that were debated in the late 1970s regarding the UK's economic performance (see Blackaby, 1979), which focused on high levels of import penetration into the UK, labor relations, low investment, short-termism, the size of the public sector, and so on. Second, in order to promote some consistency in the discussion, it is imperative that we define explicitly what "competitiveness" means. Without a clear definition, any overview of what trade theory can contribute will leave us, at best, talking at cross purposes and, at worst, saying nothing constructive at all on the competitiveness issue.

The chapter is organized as follows: The section following the introduction defines the "competitiveness" problem and identifies what factors are most likely to determine a country's competitiveness. Next we consider what, if anything, traditional trade theory can offer in understanding the competitiveness issue. A general overview of recent developments in trade theory and what insights they bring to the competitiveness debate is presented in next, followed by a discussion of the various policy options in the context of these recent theoretical developments. Finally, we summarize our conclusions.

Defining the Competitiveness Problem

While there has been a profusion of literature in recent years relating to U.S. competitiveness, it is difficult to find a useful definition of "competitiveness." This may be due to the fact that many commentators do not regard it as necessary to make explicit an appropriate definition, perhaps because it is obvious, or because "competitiveness" means different things to different people. Yet, in order to promote some consistency in our discussion, it is necessary to start with a clear definition in mind.

A concise definition of competitiveness is given by Fagerberg (1988) who defines it as:

> ... the ability of a country to realize central economic policy goals, especially growth in income and employment, without running into balance-of-payments difficulties (p. 355).

This definition is not inconsistent with others that have been found in the literature. For example, Hatsopolous *et al.* (1988) define "competitiveness" as:

> ... not simply the ability of a country to balance its trade, but its ability to do so

while achieving an acceptable improvement in its standard of living...[Further] we would not regard the United States as competitive unless it was able to maintain a rate of growth in living standards that keeps pace with that in the rest of the world (p. 299).

while from the business school camp, Scott (1985) defines it as:

...a nation's ability to produce, distribute and service goods in the international economy in competition with goods and services in other countries, and to do so is a way that earns a rising standard of living (pp. 14-15)

There are perhaps three points worth emphasizing with regard to these definitions. First, competitiveness is primarily about economic growth. Second, according to these definitions, competitiveness is not explicitly about either market share issues or other indicators of industry or sector performance. These aspects of performance are important only insofar as they relate to economic growth. Third, competitiveness is a long-run issue: Growth, by definition, is a path-dependent process such that the current allocation of resources in the economy will determine future standards of living. As McCulloch (1985) points out:

...some policies could increase market share ... but may achieve these results *at the expense of future gains* in productive capacity, employment and national well-being (p. 143) [Emphasis added].

Armed with Fagerberg's (*op.cit.*) definition, one can turn to the question as to what determines a country's competitiveness. Traditionally, discussion of the determinants of competitiveness has focused on manufacturing cost comparisons between countries (Fagerberg, *op.cit.*). Thus, for example, the debate on the UK's deteriorating economy often emphasized high relative unit labor costs as a cause of deindustrialization in the UK. The assumption here is that, for a given (constant) markup onto final good prices, if a country's unit labor costs were relatively lower vis-à-vis other countries it would gain global market share which, in turn, would be expected to increase growth. Similarly, relative export prices would also reflect cost advantages.

All this appears obvious. There is one basic problem, however: Changes in relative costs do not appear to correspond with expected changes in market shares. This was first noted by Kaldor (1978) and is sometimes referred to as the "Kaldor paradox." Specifically, Kaldor presented data on relative unit labor costs and unit export values for eleven major industrialized countries for the period 1963-1975 and compared them with changes in each country's global market share. He found that, in six of the eleven cases (which included the

United States, Japan, West Germany, and the UK) rising (falling) relative labor costs or relative export values were matched with higher (lower) market shares. Does the "Kaldor paradox" still hold? In order to answer this, data on normalized unit labor costs, unit export values, and market share were obtained for West Germany, Japan, and the United States for the period 1966-1985. Market share is defined as a country's share of total world imports. Relative export prices for each country are for manufacturing goods only, this data being drawn from the recent data set compiled by Lipsey *et al.* (1991). The data are presented in Table 7.1 and are expressed as average annual percentage changes.

Conventional wisdom would lead us to expect that lower relative wages would be reflected in higher market shares. However, for all three countries, our *a priori* convictions are not upheld: Market share appears to be positively correlated with changes in relative costs.[1] Similarly, relative export prices appear to be positively correlated with market shares for the United States and Japan, though not so for Germany. It therefore appears that the "Kaldor paradox" still holds: At least at this superficial level, relative cost data do not appear to tell us very much about a country's competitiveness.[2]

If cost comparisons do not explain changing market shares for the most successful industrialized countries, what does? Useful empirical work comes from Fagerberg (*op.cit.*). Fagerberg specifies a theoretical model that relates the determinants of trade performance (relative costs, technological progress and investment) to GDP growth and tests it using panel data for 15 OECD countries for the period 1961-1983. In general, Fagerberg's results show that factors relating to technology and investment primarily determine medium and long-run differences in growth in market share and GDP across countries. Cost differences, in accordance with the data in Table 7.1 on the "Kaldor paradox," play a more limited role in explaining a country's competitiveness.

Fagerberg's econometric results appear to accord with more casual observation of the competitive challenges facing the U.S. economy.[3] These challenges have both an internal and external dimension. Internally, the United States has faced a productivity slowdown over the post-war period in a large number of

TABLE 7.1 Relative Costs and Market Shares for the United States, Germany, and Japan: 1966-1985. (Average annual percentage changes)

Country	Relative Normalized Unit Labor Costs	Relative Export Prices	Market Share
U.S.	-1.16	-1.66	-1.72
Germany	1.72	-0.80	0.62
Japan	2.18	1.44	5.04

Sources: Labor Costs: IMF Price Statistics, various. Market Shares: IMF Price Statistics, various. Export Prices: Lipsey *et al.*, 1991.

sectors, the slowdown being particularly marked over the 1970s. The U.S. productivity slowdown has largely been associated with low investment, investment in the United States as a proportion of GDP over 1970 to 1980 being lower than in most other industrialized countries (McCorriston, 1992). Low investment in the United States can, in large part, be explained by the high cost of capital, low savings rates, and other macroeconomic phenomena.

Despite the productivity slowdown, the United States has remained at the top of the productivity league (see Baumol *et al.*, 1989). Other countries have, however, converged on U.S. productivity levels. Part of this is due to higher investment levels in other countries, though R&D performance also plays an important role, particularly given that the source of the strongest challenges to the United States have come in the high technology sectors. It is notable that R&D expenditure is now higher in Japan and Germany than in the United States (National Science Foundation, 1989). This gap in R&D expenditure is exacerbated when one looks at R&D spending on nondefense activities: Data for 1987 show that nondefense R&D expenditure as a percentage of GDP was 2.8 percent in Japan, 2.6 percent in Germany and 1.75 percent in the United States.

In sum, both general observation and econometric results based on stronger theoretical foundations appear to suggest that productivity growth (and its determinants, investment, and technological progress) is the principal factor influencing a country's competitiveness. In light of this conclusion, we now turn to a discussion of traditional trade theory and what it can offer as an insight to the competitiveness issue.

Traditional Trade Theory and Competitiveness

The simplest way to study the effects of a country's deteriorating productivity performance in the context of traditional theory is to refer to a simple Ricardian model of trade. In this model, we assume that there are two countries, home and foreign. Labor is the only factor of production with total labor endowment for each country given by L and L* for the home and foreign country respectively. The amount of labor required in the production of individual goods captures factor productivity, and the wage rate is the appropriate reward to this factor. Following Dornbusch *et al.* (1977), it is assumed that a large number of goods can be produced by both countries, the range of goods z being spread over the interval [0,1]. A country will produce a proportion of goods in this range, the extent depending on relative costs of production. As we shall see, changes in factor productivity will influence the range of goods each country will specialize in.

Relative costs of production are given by:

$$a(z)w \leq a^*(z)w^* \tag{1}$$

where $a(z)$ is the labor requirement for producing z in the home country, $a^*(z)$ is the labor requirement for producing z in the foreign country and where w and w^* represent costs of production in the home and foreign country, respectively. Re-arranging (1) we have:

$$\frac{w}{w^*} \leq A(z) \tag{2}$$

where $A(z)$ equals $a^*(z)/a(z)$ and is shown in Figure 7.1 as the $A(z)$ schedule. It is downward sloping: With lower relative wages, the home country will produce a larger range of goods.

On the demand side, exports must equal imports. $v(z)$ is the share of income spent on home goods, and $(1-v)(z)$ is the share of income spent on foreign-produced goods. Demand for exports then depends on foreigners' income (w^*L^*). Similarly, the demand for imports depends on home country's income (wL). The current account therefore balances when:

$$v(z)(w^*L^*) = (1-v)(z)(wL) \tag{3}$$

which, after rearranging we obtain:

$$\frac{w}{w^*} = \frac{v(z)}{(1-v)(z)} (L^*/L) \tag{4}$$

This relationship is represented by the upward sloping schedule B in Figure 7.1, which shows that a higher demand for home goods will be offset by higher relative wages if trade is to remain balanced. Equilibrium establishes relative wages $(w/w^*)_0$ and the range of goods each country will produce. As shown in Figure 7.1, the home country will have a comparative advantage over the range $[0, \tilde{z}\ ]$ while the foreign country will produce $[\tilde{z}\ ,1]$ goods.

It is now easy to see what happens when productivity uniformly improves in the foreign country. Productivity improvement implies $a^*(z)$ is now smaller. This shifts the $A(z)$ schedule to the left as shown in Figure 7.1. If relative wages remain unchanged, the home country will now produce only $[0, \tilde{z}_1\ ]$; relative wages, however, will fall to $(w/w^*)_1$ to offset this relative productivity deterioration leaving the country producing a smaller range of goods given by $[0, \tilde{z}_2\ ]$.

How does this analysis relate to the competitiveness debate? Note that from the Ricardian model presented above, the country still participates in trade despite its relative productivity weakness, and it still gains from trade and consumers benefit from the productivity improvements in the foreign country via terms-of-trade effects. As far as traditional theory is concerned, therefore, competitiveness would appear to be largely a nonissue. Unlike companies, countries cannot go out of business even if other countries become more productive in *all* activities: Even if a country is relatively less productive in all sectors, it will still have a comparative advantage in some activities.

In the context of the competitiveness debate, the concern with these changes must, therefore, have its source elsewhere. It may be due, for example, to the adjustment costs in running down $\bar{z}_2 - \bar{z}$ industries; but this is not different from traditional demands for protection. Perhaps more convincingly there must be a concern that there are certain attributes of the $\bar{z}_2 - \bar{z}$ industries that are in some way important to the home country that are not captured by traditional theory. Thus, what seems to be important as far as the competitiveness debate is concerned is the *mix* of industries in the home country rather than market share or profitability *per se*. This accords with McCulloch's (*op.cit.*) critique of the competitiveness debate:

> Many concerns about competitiveness are actually concerns about changes in the composition of output relative to some unspecified ideal (p. 142).

As Krugman (1991) suggests, this a more subtle view of the competitiveness issue than is generally understood.

What characteristics of certain (and, by implication, key) industries does traditional trade theory miss? There are perhaps three (related) features: First, these industries may be imperfectly competitive, generating rents that can increase national welfare over time; second, and perhaps more importantly, these industries may be R&D intensive and can thus generate spillovers for the rest of the economy; and, third, these industries may be characterized by

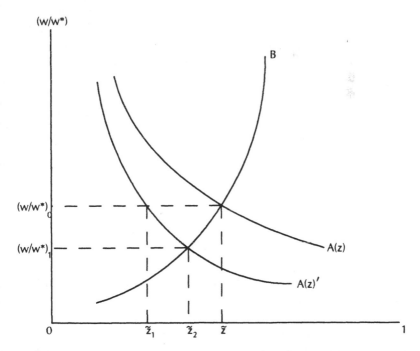

FIGURE 7.1 Relative Labor Requirements, Relative Wages, and Trade Balance.

increasing returns such that they provide inputs to other industries at decreasing costs over time. Such features are commonly associated with high technology sectors, and it appears that much of the competitiveness debate is concerned with the loss of these "strategic" sectors of the U.S. economy.

However, there is a final and perhaps more important point that traditional theory does not address. As we have seen from our definitions of competitiveness, the issue is one essentially concerned with growth. Thus, what is relevant is the dynamic (growth) effects of alternative compositions of output. Similarly, as regards policy options, it is the dynamic rather than the once-and-for-all effects of government intervention that is important. Thus, in this respect, traditional trade theory is deficient since it is the dynamic effects that are critical in understanding the competitiveness issue.

Recent Theoretical Developments

The focus of this chapter is on ascertaining how certain policy options can improve a country's competitiveness. As discussed in the previous section, the competitiveness debate is concerned with the mix of industries, particular concern being raised over the loss of key high technology sectors. Some recent developments in trade policy analysis would appear to address some of these problems since they attempt to accommodate some of the features common with these high technology sectors. This literature is often referred to as strategic trade policy analysis. The term "strategic" here is mainly associated with the role that government intervention could play in influencing the game-theoretic interaction between competing (home and foreign) firms and thus, largely, refers to the oligopolistic nature of these industries. This literature shows that government intervention can give the home firm the equivalent of first-mover advantage such that rents can be captured from foreign competitors. These theoretical developments originated with Brander and Spencer (1983, 1985), an overview of which can be found in Krugman (1987a) and Baldwin (1992). However, while much of this literature deals with the oligopolistic aspects of high technology industries, few studies have incorporated dynamic features, the notable exception being Baldwin and Krugman (1988) who incorporate learning-by-doing effects. The principal conclusion of the strategic trade policy literature has been to suggest that government intervention, through setting optimal values for import tariffs or export subsidies, can increase national welfare, although empirical studies have shown that the likely gains are small (see Helpman and Krugman, 1989).

However, given our discussion of the competitiveness problem, this literature on "rent-shifting" policy does not offer appropriate insights into how policy can influence competitiveness. First, since competitiveness is concerned with the mix of industries, a general equilibrium perspective is required. Virtually all studies of strategic trade policy are partial equilibrium in nature, although Dixit and Grossman (1986) highlight the effect of targeting one industry on other industries using similar inputs. Second, given our definitions of competitiveness, ideally the focus of policy options should be on how policy affects growth. Strategic trade policy (in common with most trade policy analysis) identifies only once-and-for-all *level* effects. In the context of the competitiveness debate, trade theory should identify dynamic aspects, i.e., the effects of policy on *rates of growth*.

Recent developments in the broad area of macroeconomic growth theory cast some light on these issues. These theoretical developments were spurred by work by Romer (1986) and Lucas (1988). The most complete and recent analysis of the interaction between growth and trade is found in Grossman and Helpman (1991b). Before discussing policy options explicitly, we outline below (albeit somewhat heuristically) some of the principal features of these recent developments that are relevant to understanding the links between policy and competitiveness.

General Equilibrium

Given that we are concerned with the mix of industries, a general equilibrium framework is required. This will give an insight into the desirability of policies that target certain sectors of the economy. Typically, in recent theoretical work, there are three sources of economic activity: a sector that produces traditional manufacturing goods; a sector producing high technology goods; and an R&D sector that produces blueprints for new goods. All sectors use both human capital and unskilled labor. The R&D sector is relatively human capital intensive while the traditional manufacturing sector uses unskilled labor relatively intensively. The high technology sector is more human capital intensive than traditional manufacturing but less so compared with the R&D sector.

Intertemporal Choice

The current allocation of resources between sectors will determine future competitiveness. Specifically, preference for higher consumption now will lead to fewer resources devoted to R&D, which will slow down the rate of innovation. Since such choices are endogenous, they must be dealt with via an intertemporal consumer utility function.

Endogenous R&D Activity

One of the most significant aspects of this recent theoretical work is the modeling of innovation as the outcome of intentional activity by entrepreneurs seeking profits.[4] Of course, this idea is not new, Schumpeter having referred to it in the 1940s.[5] Endogenizing innovative activity is an improvement on traditional growth theory where technological progress was largely treated as being exogenous. This has implications for the discussion of policy: If technological progress is exogenous, government intervention — at least from a theoretical viewpoint — cannot influence it. Now that it can be treated endogenously, policy has a potential role in influencing the growth rate.

However, in the Schumpeterian tradition, when new products are developed, monopoly pricing can arise. This creates a distortion in the economy, i.e., the volume of innovative goods available will be too low.

Characteristics of R&D

There are several features of R&D in these models that are notable. First, on a point that Romer (1990) has emphasized, R&D output is nonrival in nature. Thus, the fact that someone uses a mathematical formula or a firm utilizes an engineering design does not stop others from using it at the same time. Further, R&D output will be (at least partially) nonexcludable; there is nothing to stop others from using it. Thus, R&D plays two roles in these models: It creates new designs for innovative goods and it adds to society's stock of R&D knowledge. As Romer (*ibid.*) points out, with these features of R&D, an economy will be characterized by increasing returns to scale.

Second, technical progress in recent research usually takes one of two forms. The simplest one is where R&D adds to the number of innovative goods available. Thus, the production function used in this case is similar to Ethier's (1982) adaptation of Dixit and Stiglitz's (1977) consumer utility function. Given this feature of R&D activity, a result highlighted by Romer (1986) and Grossman and Helpman (1991b) is that private R&D leads to a suboptimal level of innovation since entrepreneurs do not account for the contribution their R&D activity makes to the economy's stock of knowledge, i.e., entrepreneurs ignore the spillover benefits when making their decision to invest in R&D activity.

To a certain extent, however, this specification is undesirable since observation informs us that goods are improved over time. This is dealt with by Grossman and Helpman (1991a) who allow for goods increasing in quality rather than in number. R&D activity is more sophisticated in this case. In seeking profits, entrepreneurs allocate resources to R&D in order to improve upon the highest quality currently available. If successful, the incumbent will cease production since consumers will prefer to buy the new higher quality good at the quality-adjusted price. However, the firm that innovates successfully knows that other firms will target R&D to upgrade this good and, in time, will make zero profit once displaced. Thus, in this model, endogenous R&D activity results in a quality ladder with goods being improved over time.

Grossman and Helpman (*ibid.*) show that, in this model, innovation can be too slow or too fast. There are three externalities of which the private entrepreneur does not take any account. First, he ignores that consumer surplus increases with the new higher quality good. Second, as before, there is the contribution that his R&D efforts make to the stock of knowledge. Third, there is the profit destruction effect that arises when the incumbent's profits fall to zero if he successfully develops a higher quality good. The first two effects are positive; the latter negative. As Grossman and Helpman (1991a) show, the outcome depends on whether the incentive to allocate resources to R&D is too high. If so, the rate of innovation will be too fast; otherwise, as in the expanding variety case, the rate of innovation will be too low.

First Best Policies

Before discussing specific policy issues, it is clear from the above overview that there is a role for government. There are two distortions: monopoly pricing in the high technology sectors and perhaps too little output emanating from the R&D sector. First best policy is, therefore, a subsidy to counter the first distortion of insufficient volume of the high technology good being available and an R&D subsidy to deal with the second. However, if there is too much R&D activity that can arise in the quality ladder model, rather than a subsidy, a tax on R&D activity should be used.

Given the general framework these theoretical developments provide, we are now in a position to deal with explicit policy issues: Will R&D subsidies, industrial policy, or trade policy improve a country's competitiveness?

Policy and Competitiveness

We refer to Grossman and Helpman's (1991b) framework for dealing with alternative policy options aimed at improving a country's competitiveness. Their framework typifies a world consisting of two countries (say A and B), both of which allocate resources between a traditional manufacturing sector, a high technology sector producing vertically differentiated goods and R&D activity. Each researcher can take advantage of R&D activity in either country, and with incomplete specialization and factor price equalization, the rate of growth is common to both countries. As we shall see, the advisability of policy will, in some cases, depend on the pattern of comparative advantage in each country. Further, we refer to the case where technological progress involves higher quality goods rather than an increasing number with the features of R&D activity as outlined above. We consider the various policy options below.

R&D Subsidies

Suppose, in line with some demands, the government in country A subsidizes R&D so that the private cost of R&D activity in country A falls. As a result of this policy, the composition of output changes in each country. The R&D sector in country A expands as a result of the subsidy and correspondingly contracts in country B, though the aggregate rate of innovation increases

(Grossman, 1989 and Grossman and Helpman, 1991b, Ch. 10). However, the foreign country now produces a larger fraction of the world's output of high technology goods. This arises since the effect of the R&D subsidy in country A encourages R&D activity, thus drawing human capital from its high technology sector. Expansion of the R&D sector in country A is, therefore, at the expense of its high technology sector. In country B, however, since R&D activity falls, human capital is released which is employed in the high technology sector.

Grossman (*op.cit.*) argues that this scenario could partially explain why Japan has captured a greater share of the world market in high technology goods. Evidence shows that government funding of R&D in Japan is lower than in other major industrialized countries. For example, in 1986, the percentage of total R&D funded by the Japanese government was 19.6 percent but was 48.3 percent in the United States. Corresponding figures for Germany, France and the UK were 37.5 percent, 46.1 percent, and 42.2 percent, respectively (Grossman, *op.cit.*). Consequently, the logic of the argument is that Japan allocates more of its skilled labor to the high technology sector and less to original R&D. In sum, an R&D subsidy would, in this scenario, increase growth in *both* countries though the subsidizing country would produce a smaller fraction of the high technology goods.

Production Subsidy

As suggested earlier, there have been demands for a more activist industrial policy in the United States (Jarboe *op.cit.*). The use of production subsidies (say, in the guise of government procurement) may be an integral part of such a policy. Would they be desirable? Such a subsidy has two effects. On the one hand, it increases the profitability of R&D in country A; on the other hand, expansion of the high technology sector will increase the cost of R&D. In the Grossman and Helpman (1991b) model, the latter effect dominates and a lower rate of innovation results. This applies to both countries despite the expansion of the R&D sector in country B. Thus, the production subsidy expands country A's high technology sector but at the expense of lower growth. By the same mechanism, a production subsidy to the traditional goods sector will increase growth. R&D costs subsequently fall and growth increases, though the high technology sector will contract in the process.

These examples highlight the importance of adopting a general equilibrium framework for understanding the effects policies may have on a country's competitiveness. What is important is the resource the targeted sector uses intensively. Since the high technology sector and R&D activity compete for a similar bundle of resources, expanding one of these sectors occurs at the expense of the other. Traditional manufacturing and R&D activity, however, appear as complements in this general equilibrium framework.

Trade Policies

Perhaps the most common feature of the current competitiveness debate is the demand for the use of trade policy instruments, either in the form of export subsidies or import tariffs, or their equivalent in other forms of trade restrictions. Again, will such policies improve competitiveness? Consider the case of an import tariff on high technology goods imposed by country A. This tariff has two component parts. On the production side, an import tariff has the same effect as a production subsidy; the expansion of high technology output raises the cost of R&D. The tariff also increases prices to consumers in country A, which lowers demand for high technology goods, which, *ceteris paribus*, lowers the cost of R&D. The net effect on growth thus depends on which of these effects dominate. If the spending effect dominates, then high technology output will fall and, on balance, the rate of innovation will increase. Thus, if a country has a comparative disadvantage in high technology goods, then the global rate of innovation will increase as a result of country A's trade policies. If, however, it has a comparative advantage in R&D, trade policy will lead to a lower growth rate.

What are the key features that arise from this discussion of policy options to improve a country's competitiveness? First, if one views competitiveness as ultimately being concerned with economic growth (see section "Defining the Competitiveness Problem"), then resources allocated to R&D worldwide are necessary to keep growth going in both countries. In this context, it is important to understand the general equilibrium aspects of targeting specific sectors in each country. Then many of the effects of government policies become intuitive. Two further aspects are worth noting: First, the effects of a country's policies can be transmitted worldwide, various policies in one country affecting the allocation of resources in both countries. This raises the question as to how one country should respond to another country's policies. This is related to the second point. With R&D spillovers between countries, each country benefits from R&D activity in its competitor country. Thus, it may be desirable for a country to remain passive even if the other country is pursuing an activist policy vis-à-vis its high technology sector.

Ultimately, however, one should be interested in the welfare implications of such policies. Unfortunately, the welfare effects are ambiguous. Since we have focussed on scenarios where R&D is characterized by quality improvements, in the steady state, there is the possibility of too much innovation. However, as long as the initial incentive to invest in R&D is not too great, welfare will increase as the rate of innovation increases. Against this, however, policies may simultaneously affect the monopolistic distortion, in some cases exacerbating it, in others not. Since the welfare effects associated with the rate of innovation and the monopolistic distortion may offset each other, it is difficult to comment unambiguously on the desirability of these policy changes on welfare grounds.

Conclusion

This chapter has offered a perspective on how alternative policy options may affect a country's competitiveness. A key feature of our discussion has been to draw upon an explicit definition of "competitiveness," which has determined the appropriate theoretical framework upon which to draw our conclusions regarding policy. Since competitiveness is ultimately concerned with growth and not specifically market share or other indicators of industry performance, the important consideration is to understand the dynamic — rather than static — effects of alternative policies, i.e., how policies influencing the current allocation of resources affect the future well-being of a country's citizens.

There are three main points that arise from our discussion. First, it is clearly important to consider the links between policy and competitiveness in a general equilibrium framework. A sectoral focus may advocate an increase in market share for a particular sector but this may occur at the expense of drawing resources away from other sectors that will ultimately reduce the country's growth rate. Second, policy intervention may be justifiable in particular sectors if the aim is to increase growth, though the welfare implications can be ambiguous. Finally, if other countries are following interventionist policies, it is not necessarily the case that other countries should respond in kind. With the results of R&D being available across national boundaries, passive countries may benefit from government intervention abroad. The introduction of policy instruments may lead to a deterioration in a country's competitiveness.

Notes

1. Of course, the direction of causality may be reversed, i.e., as a result of declining productivity, wages have to fall to maintain market share. See Hatsopolous *et al.* (1988).

2. Further support for the inadequacy of relative price data in explaining market share can be found in Kellman (1983).

3. See McCorriston (1992) for an overview of the U.S. competitiveness issue.

4. Some models focus solely on "learning-by-doing." See Lucas (1988) and Krugman (1987b).

5. For an alternative treatment of Schumpeterian R&D activity, see Krugman (1990).

References

Baldwin, R. "Are Economists' Traditional Trade Policy Views Still Valid?" *Journal of Economic Literature* 30(1992): 804-830.

Baldwin, R., and P.R. Krugman. "Market Access and International Competition: A Simulation Study of 16K Random Access Memories." In *Empirical Methods for International Trade*, edited by R. Feenstra. Cambridge, MA: The MIT Press, 1988.

Baumol, W J , S A B Blackman, and F.N. Wolff. *Productivity and American Leadership*. Cambridge, MA: The MIT Press, 1989.

Blackaby, F. *De-Industrialization*. London: Heinemann, 1979.

Brander, J.A., and B.J. Spencer. "International R&D Rivalry and Industrial Strategy." *Review of Economic Studies* 50(1983): 707-22.

———. "Export Subsidies and International Market Share Rivalry." *Journal of International Economics* 18(1985): 83-100.

Dixit, A., and G.M. Grossman. "Targeted Export Promotion with Several Oligopolistic Industries." *Journal of International Economics* 21(1986): 233-50.

Dixit, A., and J.E. Stiglitz. "Monopolistic Competition and Optimum Product Diversity." *American Economics Review* 67(1977): 297-308.

Dornbusch, R. "Policy Options for Freer Trade: The Case for Bilateralism." In *An American Trade Strategy: Options for the 1990's*, edited by R.A. Lawrence and C.L. Schultze. Washington: Brookings Institution, 1990.

Dornbusch, R., S. Fisher, and P.A. Samuelson. "Comparative Advantage, Trade and Payments in a Ricardian Model with a Continuum of Goods." *American Economic Review* 67(1977): 823-39.

Ethier, W. "National and International Returns to Scale in the Modern Theory of International Trade." *American Economics Review*, 72(1982): 389-405.

Fagerberg, J. "International Competitiveness." *Economic Journal*, 98(1988): 355-374.

Grossman, G.M. "Explaining Japan's Innovation and Trade: A Model of Quality Competition and Dynamic Comparative Advantage." NBER Working Paper, No. 3194, 1989.

Grossman, G.M., and E. Helpman. "Quality Ladders in the Theory of Growth." *Review of Economic Studies* 58(1991a): 43-61.

———. *Innovation and Growth in the Global Economy*. Cambridge, MA: The MIT Press, 1991b.

Hatsopolous, G.N., P.R. Krugman, and L.H. Summers. "U.S. Competitiveness: Beyond the Trade Deficit." *Science* 9 (July 1988): 299-307.

Helpman, E., and P.R. Krugman. *Trade Policy and Market Structure*. Cambridge, MA: The MIT Press, 1989.

Jarboe, K.N. "A Reader's Guide to the Industrial Policy Debate." *California Management Review* 27(1985): 198-220.

Kaldor, N. "The Effect of Devaluations or Trade in Manufacturers." In *Future Essays in Applied Economics*, London: Duckworth, 1978.

Kellman, M. "Relative Prices and International Competitiveness: An Empirical Investigation." *Empirical Economics* 8(1983): 125-39.

Krugman, P.R. "Is Free Trade Passé?" *Journal of Economic Perspectives* 1(1987a): 131-144.

———. "The Narrow Moving Band, the Dutch Disease, and the Competitive Consequences of Mrs. Thatcher: Notes on Trade in the Presence of Scale Dynamic Economies." *Journal of Development Economics* 27(1987b): 41-55.

———. "A Model of Innovation, Technology Transfer, and the World Distribution of Income." In *Rethinking International Trade*, edited by P.R. Krugman. Cambridge, MA: The MIT Press, 1990.

———. "Myths and Realities of U.S. Competitiveness." *Science* 12(November, 1991): 811-15.

Lipsey, R.E., L. Molnari, and I.B. Kravis. "Measures of Prices and Price Competitiveness in International Trade in Manufactured Goods." In *International Economic Transactions: Issues in Measurement and Empirical Research*, edited by P. Hooper and J.D. Richardson. Chicago: NBER, 1991.

Lucas, R.E. "On the Mechanics of Economic Development." *Journal of Monetary Economics* 22(1988): 3-42.

McCorriston, S. "An Overview of the U.S. Competitiveness Debate." *Organization and Performance of World Food Systems*. Occasional Paper, OP-43, 1992.

McCulloch, R. "Trade Deficits and International Competitiveness." *California Management Review* 27(1985): 140-57.

National Science Foundation. *Science and Engineering Indicators—1989*. Washington, D.C.:NSF, 1989.

Romer, P.M. "Increasing Returns and Long-Run Growth." *Journal of Political Economy* 94(1986): 1002-37.

———. "Are Nonconvexities Important for Understanding Growth?" *American Economic Review* 80(1990): 97-103.

Scott, B.R. "U.S. Competitiveness: Concepts, Performance and Implications." In *U.S. Competitiveness in the World Economy*, edited by B.R. Scott and G.C. Lodge. Boston, MA: Harvard Business School Press, 1985.

Tyson, L. "Managed Trade: Making the Best of Second-Best." In *An American Trade Strategy: Options for the 1990s*, edited by R.Z. Lawrence and C.L. Schultz. Washington, D.C.: Brookings Institution, 1990.

8

Technical Progress, Capital Formation, and Growth of Productivity

Lawrence J. Lau

Introduction

One important determinant of competitiveness among nations is their relative productive efficiencies. The objective of this paper is to explore the empirical relationship between capital accumulation and productivity at the aggregate national level. By productivity we mean the ability to produce real output with *given* quantities of inputs (holding the quality of the output and the inputs constant).[1] In order to measure the growth in productivity thus defined, it is necessary to estimate the increases in real output attributable to increases in the capital and labor inputs first. The percentage increase in output after the deduction of the increases in output due to the inputs may be identified as the growth in productivity (sometimes also referred to as total factor productivity) or technical progress. The three principal sources of economic growth of nations are therefore capital, labor and technical progress.

The rate of growth of labor is generally constrained by the rate of growth of population. For industrialized countries, the rate of growth of the labor force is seldom higher than two percent per annum, even with international migration. Consequently, the rate of growth of capital (physical and human) and technical progress have been found to account for a significant proportion of economic growth by a long line of distinguished economists. The early growth accounting studies — Abramovitz (1956), Solow (1957), Kendrick (1961, 1973), and Kuznets(1965, 1966, 1971, 1973) — attribute approximately one-half of the growth in output to technical progress and one-quarter to the growth in capital input. Subsequent studies, notably those by Denison, (1962, 1967, 1979, 1985), Griliches (1966), and Jorgenson (1967, 1972) and his associates (Jorgenson, Gollop, and Fraumeni, 1987) that attempt to adjust the capital and labor inputs for changes in quality, attribute between 10 and 32 percent of output growth to technical progress.

The importance of the contributions of capital and technical progress to the growth of aggregate real output can be readily understood with the help of some simple arithmetic. Starting with an aggregate production function:

$$Y_t = F(K_t, L_t, t) \tag{1}$$

where Y_t, K_t, and L_t are the quantities of aggregate real output, physical capital, and labor respectively at time t, and t is an index of chronological time, the rate of growth of output can be expressed in the familiar equation of growth accounting by taking natural logarithms of both sides of equation (1) and differentiating it totally with respect to t:

$$\frac{d\ell n Y_t}{dt} = \frac{\partial \ell n F}{\partial \ell n K}(K_t, L_t, t)\frac{d\ell n K_t}{dt} + \frac{\partial \ell n F}{\partial \ell n L}(K_t, L_t, t)\frac{d\ell n L_t}{dt} + \frac{\partial \ell n F}{\partial t}(K_t, L_t, t), \tag{2}$$

where $\dfrac{d\ell n t}{dt}$, $\dfrac{d\ell n K_t}{dt}$ and $\dfrac{d\ell n L_t}{dt}$ are the instantaneous proportional rates

of change of the quantities of real output, capital, and labor respectively at time

t; $\dfrac{\partial \ell n F}{\partial \ell n K}$ and $\dfrac{\partial \ell n F}{\partial \ell n L}$ are the elasticities of real output with respect to capital and

labor respectively at time t; and $\dfrac{\partial \ell n F}{\partial t}$ is the instantaneous rate of growth of

output holding the inputs constant, or equivalently, the rate of technical progress. The three terms on the right-hand side of equation (2) may be identified as the contribution of capital, labor and technical progress respectively to the growth in output.[2]

The production elasticity of output with respect to *measured* labor input can typically be estimated as approximately 0.6 for industrialized countries. Thus, given the rate of growth of measured labor force, which is typically no higher than 2 percent per annum in industrialized countries, the maximum rate of growth that can be accounted for by the growth in labor input is on the order of 1.2 percent. Any growth in output in excess of 1.2 percent per annum in an industrialized country is attributable to the growth in the capital input and to technical progress. For an industrialized country that grows at 3 percent per annum, approximately 60 percent of the growth in output may be attributed to physical capital and technical progress. In the short and intermediate runs, physical capital is even more important for another reason—it is the only input that can be readily varied. Human capital and technical progress can be influenced only in the longer run.

Most aggregate production function and growth-accounting studies[3] have been conducted under one or more of the traditionally maintained hypotheses of constant returns to scale in capital and labor,[4] neutrality of technical progress, and profit maximization with competitive output and input markets. The validity (or lack thereof) of each of these hypotheses affects the measurement of technical progress and the decomposition of economic growth into its sources. In this chapter, new alternative estimates of technical progress as well as the corresponding new accounts of growth, derived without maintaining these assumptions, are presented. In the next section, a new approach for the measurement of technical progress and productivity, first employed by Boskin and Lau (1990), based on the direct econometric estimation of an aggregate meta-production function, is introduced.[5] The next section contains a very brief discussion of the data and the statistical model used. Next, the major findings of Boskin and Lau (1990) are summarized. Alternative measurements of technical progress and accounts of growth based on the aggregate meta-production function estimated in Boskin and Lau (1990) are presented next and compared with those of the conventional approach. In the next section, an international and intertemporal comparison of productivity is undertaken followed by the concluding remarks that summarize the chapter.

The Meta-Production Function Approach

In Boskin and Lau (1990), a new approach to the empirical analysis of productivity and technical progress, based on the direct econometric estimation of an aggregate meta-production function, that does not require the traditionally maintained assumptions of constant returns to scale, neutrality of technical progress, and profit maximization with competitive output and factor markets, is introduced and implemented.[6] The basic assumptions for the new approach are:

(1) All countries have access to the same technology, that is, they have the same underlying aggregate production function F(.), sometimes referred to as a meta-production function, but may operate on different parts of it. The production function, however, applies to standardized, or "efficiency-equivalent," quantities of outputs and inputs, that is:

$$Y^*_{it} = F(K^*_{it}, L^*_{it}) \ , \ i = 1, \ldots, n \ ; \tag{3}$$

where Y^*_{it}, K^*_{it} and L^*_{it} are the "efficiency-equivalent" quantities of output, capital, and labor respectively of the ith country at time t, and n is the number of countries.

(2) There are differences in the technical efficiencies of production and in the qualities and possibly definitions of measured inputs across countries. However, in general, the "efficiency-equivalent" quantities of output and inputs of each country are not directly observable. It is assumed that the measured outputs and inputs of the different countries may be converted into standard-

ized, or "efficiency-equivalent," units of outputs and inputs by multiplicative country- and output- and input-specific time-varying augmentation factors, $A_{ij}(t)$'s, $i = 1, \ldots, n$; $j = 0, K, L$.[7]

$$Y^*_{it} = A_{i0}(t)Y_{it} \; ; \quad K^*_{it} = A_{iK}(t)K_{it} \; ; \quad L^*_{it} = A_{iL}(t)L_{it} \; ; i = 1, \ldots, n.[8] \qquad (4)$$

These two assumptions together imply that the aggregate production function is the same in all countries in terms of "efficiency-equivalent" units of outputs and inputs. In terms of the *measured* quantities of outputs, the production function may be rewritten as:

$$Y_{it} = Ai_0(t)^{-1}F(K^*_{it}, L^*_{it}) \; , \; i = 1, \ldots, n \; ; \qquad (5)$$

so that the reciprocal of the output-augmentation factor $A_{i0}(t)$ has the interpretation of the possibly time-varying level of the technical efficiency of production, also referred to as output efficiency, in the ith country at time t. In the empirical implementation, the commodity augmentation factors are assumed to have the constant exponential form with respect to time. Thus:

$$Y^*_{it} = A_{i0} \exp(c_{i0} \, t)Y_{it} \; ; \; K^*_{it} = A_{iK} \exp(c_{iK} \, t)K_{it} \; ;$$

$$\text{and } L^*_{it} = A_{iL} \exp(c_{iL} \, t)L_{it} \; ; \; i = 1, \ldots, n; \qquad (6)$$

where the A_{i0}'s, A_{ij}'s, c_{i0}'s, and c_{ij}'s are constants. We shall refer to the A_{i0}'s and A_{ij}'s as *augmentation level* parameters and c_{i0}'s and c_{ij}'s as *augmentation rate* parameters. For at least one country, say the ith, the constants A_{i0} and A_{ij}'s can be set identically at unity (or some other arbitrary constants), reflecting the fact that "efficiency-equivalent" outputs and inputs can be measured only relative to some standard. Econometrically this means that the constants A_{i0}'s and A_{ij}'s cannot be uniquely identified without some normalization. Without loss of generality we take the A_{i0} and A_{ij}'s for the United States to be identically unity. Subject to such a normalization, it turns out that these commodity augmentation level and rate parameters can in fact be estimated simultaneously with the parameters of the aggregate production function from pooled intercountry time-series data on the quantities of *measured* outputs and inputs. There is thus no need to rely on arbitrary assumptions. It is actually possible to answer the question of how many units of labor in country B is equivalent to 1 unit of labor in country A at some given time t empirically.

(3) The wide ranges of variation of the inputs resulting from the use of intercountry time-series data necessitate the use of a flexible functional form for F(.) above. In Boskin and Lau (1990), the aggregate meta-production function is specified to be the transcendental logarithmic (translog) functional form introduced by Christensen, Jorgenson, and Lau (1973). For a production

function with two inputs, capital (K) and labor (L), the translog production function, in terms of "efficiency-equivalent" output and inputs, takes the form

$$\ln Y^*_{it} = \ln Y_0 + a_K \ln K^*_{it} + a_L \ln L^*_{it} + B_{KK}(\ln K^*_{it})^2/2 + B_{LL}(\ln L^*_{it})^2/2$$

$$+ B_{KL} \ln K^*_{it})(\ln L^*_{it}), \, i = 1, \ldots, n. \tag{7}$$

By substituting equations (6) into equation (7), and simplifying, we obtain equation (8), which is written entirely in terms of observable variables:

$$\ln Y_{it} = \ln Y_0 + \ln A^*_{i0} + a^*_{iK} \ln K_{it} + a^*_{iL} \ln L_{it}$$

$$+ B_{KK}(\ln K_{it})^2/2 + B_{LL}(\ln L_{it})^2/2 + B_{KL}(\ln K_{it})(\ln L_{it})$$

$$+ c^*_{i0t} + (B_{KK} c_{iK} + B_{KL} c_{iL})(\ln K_{it})t + (B_{KL} c_{iK} + B_{LL} c_{iL})(\ln L_{it})t$$

$$+ (B_{KK}(c_{iK})^2 + B_{LL}(c_{iL})^2 + 2B_{KL}c_{iK}c_{iL})t^2/2, \, i = 1, \ldots, n, \tag{8}$$

where $A^*_{i0}, a^*_{iK}, a^*_{iL}$ and c^*_{i0} are country-specific constants. We note that the parameters B_{KK}, B_{KL} and B_{LL} are independent of i, i.e., of the particular individual country. They must therefore be identical across countries — thus supplying the common link among the aggregate production functions of the different countries. This also provides a basis for testing the first maintained hypothesis of this study, namely, that there is a single aggregate meta-production function for all the countries. We note further that the parameter corresponding to the $t^2/2$ term for each country is not independent but is completely determined given $B_{KK}, B_{KL}, B_{LL}, c_{iK}$ and c_{iL}. This provides a basis for testing the second maintained hypothesis of the study, namely, that technical progress may be represented in the constant exponential commodity-augmentation form.

Equation (8) is the most general specification possible under our maintained hypotheses of a single meta-production function and constant exponential commodity-augmentation representation of technical progress. Conditional on the validity of equation (8), the traditionally maintained hypotheses of growth accounting—constant returns to scale, neutrality of technical progress, and profit maximization with competitive output and input markets can be tested.[9]

In addition to the aggregate meta-production function, we also consider the behavior of the share of labor costs in the value of output: $w_{it}L_{it}/p_{it}Y_{it}$, where w_{it} is the nominal wage rate and p_{it} is the nominal price of output in the ith country at time t, as a function of measured capital and labor inputs:

$$\frac{w_{it}L_{it}}{p_{it}Y_{it}} = \frac{\partial \ln Y_{it}}{\partial \ln L_{it}}$$

$$= a^*_{iLi} + B_{KLi} \, \ell n \, K_{it} + B_{LLi} \, \ell n \, L_{it} + B_{iLt} \, t, \, i = 1, \ldots, n. \qquad (9)$$

Equations (8) and (9) constitute the estimating equations for this study. Under competitive output and input markets, the assumption of profit maximization with respect to labor, which is a necessary condition for overall profit maximization, implies that the elasticity of output with respect to labor is equal to the share of labor cost in the value of output. In other words, the parameters in equation (9) are identical to the corresponding ones in equation (8). Thus,

$$\frac{w_{it}L_{it}}{p_{it}Y_{it}} = a^*_{iL} + B_{KL} \, \ell n \, K_{it} + B_{LL} \, \ell n \, L_{it} + (B_{KL} \, c_{iK} + B_{LL} \, c_{iL})t, \, i = 1, \ldots, n. (10)$$

This provides a basis for testing the hypothesis of profit maximization with respect to labor.

The Data and the Statistical Model

Boskin and Lau (1990) use data from the Group-of-Five (G-5) countries: France, West Germany, Japan, the United Kingdom, and the United States. The period covered is from 1957 to 1985 except for West Germany and the United States, data for which begin in 1960 and 1948 respectively. The aggregate real output of each country is measured as the real Gross Domestic Product (GDP) in 1980 prices. Labor is measured as the number of person-hours worked. The share of labor in the value of output is estimated by dividing the current labor income (compensation of employees paid by resident producers) by the current GDP of each country. Capital is measured as utilized capital — the private nonresidential gross capital stock multiplied by the rate of capacity utilization. These data are converted into U.S. dollars using 1980 exchange rates. Time is measured in years chronologically with the year 1970 being set equal to zero. A detailed explanation of the variables and the data sources is given in Boskin and Lau (1990).

We introduce stochastic disturbance terms e_{1it}'s and e_{2it}'s into the first-differenced forms of the natural logarithm of the aggregate production function and the labor share equation, respectively.[10] We assume:

$$E\begin{bmatrix} \varepsilon_{1it} \\ \varepsilon_{2it} \end{bmatrix} = 0, \forall i, t; \text{ and} \qquad (11)$$

$$E\begin{bmatrix} \varepsilon_{1it} \\ \varepsilon_{2it} \end{bmatrix} = \Sigma, \text{ a constant, non-singular matrix, } \forall i, t; \qquad (12)$$

and the stochastic disturbance terms are uncorrelated across countries and over time. In the first-differenced form, our stochastic assumptions amount to saying that the influence of the stochastic disturbance terms is permanent —

they raise or lower the production function and the labor share permanently until further changes caused by future stochastic disturbance terms.

Under the further assumption of joint normality of the stochastic disturbance terms, the system of two equations consisting of the production function and the labor share equation, first differenced, and its various specializations under the different null hypotheses are estimated by the method of nonlinear instrumental variables.[11] The list of instrumental variables used in the estimation is given in Boskin and Lau (1990).

Summary of the Major Findings

The major findings of Boskin and Lau (1990) are that (1) the traditionally maintained hypotheses of constant returns to scale in capital and labor, neutrality of technical progress and profit maximization with competitive output and input markets can all be rejected, (2) in the post-war period technical progress can be represented in the purely capital-augmenting form rather than the often assumed neutral or labor-augmenting form; (3) the elasticity of output with respect to measured capital input is much lower than the usual factor-share estimate based on the assumptions of constant returns to scale and profit maximization; and (4) returns to scale are not fixed but variable and have been decreasing rather than constant. All of these findings have implications for the measurement of technical progress and growth accounting. We discuss these findings in turn.

Tests of Hypotheses[12]

First, the maintained hypotheses of the meta-production function approach adopted by Boskin and Lau (1990) are tested. They consist of (1) the aggregate production functions of all five countries are identical in terms of "efficiency-equivalent" inputs, that is, there is a single aggregate meta-production function for all countries; and (2) technical progress can be represented in the commodity-augmentation form with each augmentation factor being an exponential function of time, conditional on the single meta-production function hypothesis. Neither hypotheses can be rejected at any level of significance. The non-rejection of these two maintained hypotheses lends empirical support to the validity of the aggregate meta-production function with commodity augmentation factors approach.

Next, the three major hypotheses traditionally maintained for aggregate production function or growth accounting studies — constant returns to scale, neutrality of technical progress and profit maximization with competitive output and input markets — conditional on the validity of our maintained hypotheses of a single meta-production function and exponential commodity augmentation factors are separately tested. It is found that all of these hypotheses can be rejected at their assigned levels of significance. The results of this series of tests suggest that the traditional assumptions are not valid, at least not for the countries and time periods under study.

After establishing the validity of the assumptions of the meta-production function approach and the lack of validity of the traditional assumptions, Boskin and Lau (1990) proceed to examine the nature of technical progress. Specifically, it is found that the hypotheses of identical capital and labor augmentation level parameters across countries cannot be rejected. This implies that in the base year (1970), the "efficiencies" of measured capital and labor were not significantly different across countries.[13] It is also found that the null hypothesis that technical progress can be represented by a single (instead of three) set of augmentation rate parameters for capital cannot be rejected. Technical progress may thus be identified as purely *capital-augmenting*, i.e., Solow-neutral, rather than labor-augmenting (Harrod-neutral) or output-augmenting (Hicks-neutral), as is more frequently assumed. Thus, the aggregate production functions may be written in the form:

$$Y_{it} = A_{i0}^{-1} F(\exp(c_{iK}t)K_{it}, L_{it}), i = 1, \ldots, n. \tag{13}$$

It is apparent from equation (13) that technical progress and capital are complementary to each other.

A final hypothesis on the nature of technical progress is that of identical capital augmentation rate parameters across countries.[14] This hypothesis, conditional on the maintained hypotheses of the study, and the hypotheses of identical capital and labor augmentation level parameters, and purely capital-augmenting technical progress, can be rejected. In fact, the five countries fall into two groups: France, West Germany, and Japan all have capital augmentation rates in the range of 11-13 percent per annum whereas the U.K. and the United States have capital augmentation rates of approximately 7 percent per annum.

The Estimated Aggregate Meta-Production Function

Boskin and Lau (1990) synthesize the results of the hypothesis testing and impose the restrictions implied by the hypotheses that are not rejected at the assigned levels of significance, namely, identical augmentation level parameters for capital and labor and zero augmentation rate parameters for output and labor. The estimation results are presented in Table 8.1. The estimated capital augmentation rate parameters are statistically significant and positive for all countries. Japan has the highest rate—12.8 percent per annum, followed by France (12.1 percent), West Germany (11.0 percent), and the United Kingdom (7.2 percent). The United States has the lowest rate — 6.9 percent per annum.

TABLE 8.1 Estimated Parameters of the Aggregate Production Function and the Labor Share Equation (First-Differenced Form)

Parameter	Estimate	T-ratio
Aggregate Production Function		
a_K	0.178	4.430
a_L	0.521	1.974
B_{KK}	-0.056	-3.975
B_{LL}	-0.001	-0.004
B_{KL}	0.030	1.690
c_{FK}	0.121	5.733
c_{GK}	0.110	5.314
c_{JK}	0.128	4.432
c_{UKK}	0.072	4.922
c_{USK}	0.069	5.663
$\bar{R}^2$	0.855	
D.W.	1.818	
Labor Share Equation		
B_{KLF}	-0.145	-2.030
B_{KLG}	-0.164	-1.653
B_{KLJ}	-0.061	1.647
B_{KLUK}	0.070	0.503
B_{KLUS}	-0.028	-0.449
B_{LLF}	0.148	0.687
B_{LLG}	0.261	1.689
B_{LLJ}	-0.399	-4.317
B_{LLUK}	0.210	1.741
B_{LLUS}	0.090	0.572
B_{FLt}	0.010	2.561
B_{GLt}	0.011	2.043
B_{JLt}	0.013	3.267
B_{UKLt}	-0.002	-0.467
B_{USLt}	0.001	0.432
$\bar{R}^2$	0.188	
D.W.	1.797	

Production Elasticities

The estimates of some parameters of interest are presented in Table 8.2. The estimated production elasticities of labor range between 0.50 for Japan and 0.55 for the United States. For the year 1970, our estimates of the production elasticities of labor are somewhat higher than the actual shares of labor costs in total output for France and Japan and somewhat lower for West Germany, the United Kingdom, and the United States. The estimates of the production

TABLE 8.2 Estimated Parameters of the Aggregate Production Functions (at
1970 Values of the Independent Variables)

	Capital Elasticity	Labor Elasticity	Degrees of Local Returns to Scale	Rate of Local Technical Progress	Actual Labor Share
France	0.251 (8.241)	0.506 (5.945)	0.758 (9.565)	0.030 (12.937)	0.489
W. Germany	0.234 (9.395)	0.519 (7.792)	0.752 (11.944)	0.026 (9.299)	0.532
Japan	0.292 (8.056)	0.498 (6.399)	0.791 (9.752)	0.037 (9.012)	0.435
U.K.	0.242 (8.884)	0.513 (7.111)	0.755 (11.138)	0.017 (7.753)	0.597
U.S.	0.205 (10.426)	0.548 (5.766)	0.753 (8.189)	0.014 (7.070)	0.614

Note: Numbers in parentheses are t-ratios.

elasticities of capital range between 0.20 for the United States and 0.29 for Japan
in 1970 and are much lower than those estimated by the more conventional
factor-share method under the assumptions of constant returns to scale and
profit maximization with competitive markets. Given the values of the actual
labor shares, the factor-share estimates of the production elasticities of capital
would have been between 0.4 and 0.5, approximately twice our estimated
production elasticities of capital.

Returns to Scale

At the 1970 values of the measured inputs of each country, statistically
significant decreasing returns to scale are found for all five countries. The
estimated local degrees of returns to scale range between approximately 0.75
and 0.79. The finding of decreasing returns to scale may possibly be attributed
to omitted factors of production such as land, public capital stock, human
capital, R&D capital stock, natural resources, and the environment.

Alternative Estimates of Technical Progress
and Accounts of Growth

Alternative Measurements of Technical Progress

Our first application of the estimated aggregate meta-production function is
to use it to derive alternative estimates of the average annual rates of technical
progress, without relying on the assumptions of constant returns to scale,
neutrality of technical progress, and profit maximization. In Table 8.3, the
estimates of the average annual rates of technical progress based on the

TABLE 8.3 Alternative Estimates of Average Annual Rates of Technical Progress

Country	Conventional Estimates	Our Estimates
France	.019	.029
W. Germany	.013	.024
Japan	.016	.038
U.K.	.012	.017
U.S.	.007	.015

estimated aggregate meta-production function of Boskin and Lau (1990)[15] are presented and compared with estimates obtained with the same data but using the conventional change in total factor productivity formula.[16] There are significant differences between the two alternative sets of estimates of technical progress — the Boskin and Lau (1990) estimates are much higher, partially reflecting their finding of a lower production elasticity of capital and hence decreasing returns to scale for the five countries. However, the rankings of the countries by the average annual rate of (realized) technical progress change only marginally, with France and Japan trading first and second places.

Alternative Accounts of Economic Growth

A second application of the estimated aggregate meta-production function is to use it to assess the relative contributions of the three sources of growth — capital, labor, and technical progress — again without relying on the assumptions of constant returns to scale, neutrality of technical progress and profit maximization. In Table 8.4, we present two alternative sets of estimates of the relative contributions of the different sources of growth for each of the five

TABLE 8.4 Relative Contributions of the Sources of Growth

	Capital	Labor	Technical Progress
	This Study		
France	28	-4	76
W. Germany	32	-10	78
Japan	40	5	55
UK	32	-5	73
US	24	27	49
	Conventional Estimates		
France	56	-4	48
W. Germany	69	-11	42
Japan	72	5	23
UK	55	-6	51
US	47	30	23

countries, first using the estimated aggregate meta-production function of Boskin and Lau (1990) and second using the conventional change in total factor productivity formula, again based on the same data. Table 8.4 shows, according to the meta-production function, that over the period under study, technical progress is by far the most important source of economic growth, accounting for more than 49 percent (more than 70 percent for the European countries), and capital is the second most important source of economic growth (except for the United States). Labor accounts for less than 5 percent except for the United States. These results may be contrasted to those of the conventional approach which identify capital as the most important source of economic growth (more than 47 percent), followed by technical progress (between 23 and 51 percent).

It is interesting to note that the estimated *combined* contributions of capital and technical progress are virtually the same under both approaches as are the contributions due to labor. By either the new or the conventional method, capital and technical progress combined account for more than 95 percent of the economic growth of France, West Germany, Japan, and the United Kingdom. In the United States, where the labor force grew more rapidly than in other countries during this period, they still account for more than 70 percent of the economic growth. Thus, despite the differences in the underlying assumptions, the estimates of the combined contributions are not qualitatively different.

The reason the estimates of the combined contributions of capital and technical progress are so similar is that the estimated contributions of labor are very similar by either approach—our estimated output elasticities with respect to labor are not that different from those obtained by the factor share method. Thus, the estimated combined contributions of the other factors — capital and technical progress—must also be very similar. However, our approach yields much lower output elasticities with respect to capital than those obtained by the factor share method under the constant returns to scale assumption. Thus, our estimated contributions of the remaining factor, technical progress, or the "residual," must be correspondingly higher. Another way of understanding our results is to observe that our low estimated production elasticities of capital imply decreasing rather than constant returns to scale and thus the estimated rates of technical progress must be higher to be consistent with the same rates of growth of real output and inputs.

International and Intertemporal Comparison of Productivity[17]

As a further application, we compare the evolution of the productive efficiencies of the different countries over time. In Figure 8.1, we plot the real output per measured labor-hour of each of the five countries against time. The United States had the highest real output per labor-hour until it was overtaken by France and West Germany in the late 1970s. The United Kingdom fell behind

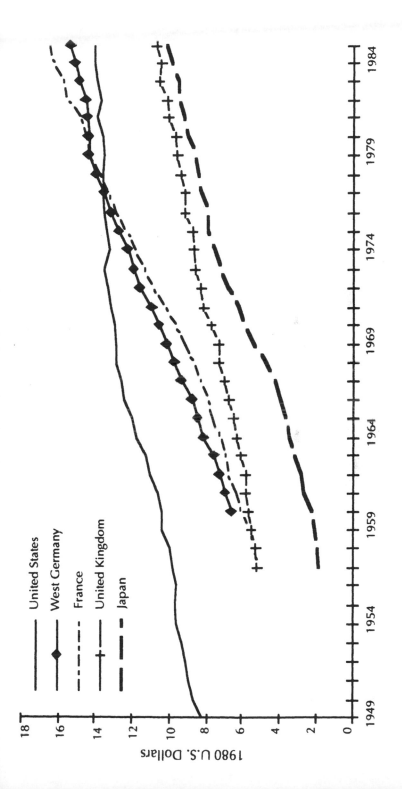

FIGURE 8.1 Real Output per Labor Hour.

France and West Germany in the early 1960s. Japan started in the last place at a very low level but by 1985 had narrowed the gap considerably. However, real output per labor-hour may differ across countries because of differences in capital intensity (capital stock per unit labor) and scale, as well as in efficiency and technical progress, not to mention the exchange rates used to convert the real outputs to a comparable basis. In Figure 8.2, we plot the quantity of the real capital stock per worker in the labor force of each of the five countries against time. We note that the United States had the highest level of capital stock per worker until around 1970, when it was overtaken by the European countries, due in part to the higher rate of growth of the labor force in the United States. However, the measured capital stock per worker of the United States was still significantly higher than that of Japan as of 1985 even though the rate of growth of the Japanese capital stock was almost three times that of the United States in the postwar period. As of 1985, West Germany had the highest measured capital stock per worker, followed by France and the U.K.

In order to compare overall productive efficiencies across countries, we must net out the effects of differing quantities of inputs (capital intensity and scale). We therefore pose the hypothetical question: If all countries have the same quantities of measured inputs of capital and labor as the United States, but their own augmentation factors (that is, efficiencies), what would have been the quantities of their real outputs and how would they evolve over time? In other words, we compare their abilities to produce real output, holding *measured* inputs constant.

To answer this question we project the time series of hypothetical real outputs for each country by the formula:

$$\ln\hat{Y}_{it} = \ln Y_0 - \ln A_{i0} + a_K \ln K_{USt} + a_L \ln L_{USt} + B_{KK}(\ln K_{USt})2/2$$

$$+ B_{LL}(\ln L_{USt})2/2 + B_{KL}(\ln K_{USt})(\ln L_{USt}) + (a_K c_{iK})t + (B_{KK}c_{iK})(\ln K_{USt})t$$

$$+ (B_{KL}c_{iK})\ln L_{USt})t + (B_{KK}(c_{iK})^2)t^2/2, \, i = 1,\ldots,n, \tag{14}$$

substituting in the estimated values of the parameters.[18] Equation (14) gives the level of real output that would have been produced by each country in each period if it had the measured inputs of the United States in that period. The results are plotted for each country in Figure 8.3. Figure 8.3 shows that in 1949 the United States had the highest level of overall productive efficiency, and West Germany the second highest (but considerably lower than the United States), and Japan the lowest. By the late 1950s France and Japan had overtaken the United Kingdom. As of 1985, the United States remained in the first place as the most productive economy (in terms of having the highest output given the same measured inputs) and the United Kingdom in last place, with West

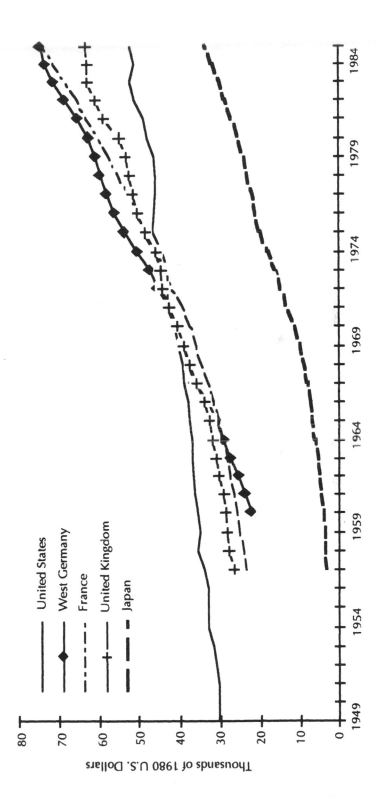

FIGURE 8.2 Capital Stock per Worker.

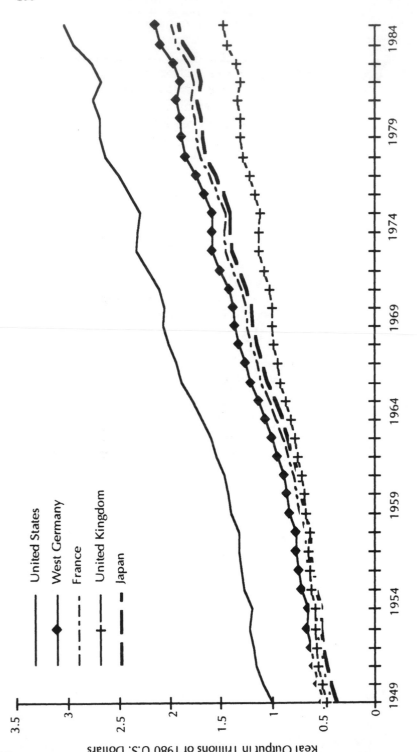

FIGURE 8.3 Hypothetical Outputs of the Group-of-Five Countries with U.S. Measured Inputs.

Germany, France, and Japan closely clustered together. However, U.S. productivity has not been growing as fast as those of some of the other countries and thus its "advantage" has been declining. With the same measured inputs, Japan could produce less than 40 percent of United States' aggregate real output in 1949, but almost two-thirds of United States' aggregate real output in 1985. Similar improvements were made by France and West Germany. However, the gap between the United Kingdom and the United States narrowed only very slightly during this period. In Figure 8.4 we plot the relative productive efficiency of each of the four countries against time, using the United States level as the reference (that is, with U.S. productive efficiency normalized at unity). Figure 8.4 provides the same picture as Figure 8.3, namely, that France, West Germany, and Japan have closed the gap significantly but not the United Kingdom.

Concluding Remarks

We have introduced and implemented a new method of analyzing technical progress and economic growth based on the concept of an aggregate meta-production function, using pooled time series data from the Group-of-Five countries for the postwar period. We have found that the empirical data are inconsistent with the traditionally maintained hypotheses of constant returns to scale, neutrality of technical progress, and profit maximization under competitive conditions, at the aggregate, national level. Instead, we find purely capital-augmenting technical progress as well as sharply decreasing local returns to scale. All the traditional hypotheses are, however, customarily maintained in the application of the conventional method of measuring the rate of technical progress and of growth accounting.

With our new approach, we have obtained alternative estimates of the rates of technical progress as well as alternative decompositions of economic growth into its sources — capital, labor and technical progress — that are independent of the conventional assumptions. We have found much higher rates of *realized* technical progress. We have also found that technical progress is by far the most important source of economic growth of the industrialized countries in our sample, accounting for more than 49 percent, in contrast to the growth accounting studies of Denison (1962, 1967, 1979, 1985) and Griliches and Jorgenson and his associates. However, we have not made explicit adjustments for the quality of capital or labor, as were done by Denison and Jorgenson, Gollop and Fraumeni (1987). Instead, we allow any trend of improving input quality to be captured by the rates of capital and labor augmentation themselves. Thus, what we attribute to technical progress may include what others attribute to the improvement in the qualities of the inputs.

We have also found that technical progress is of the capital-augmenting rather than the more frequently assumed output-augmenting (Hicks-neutral)

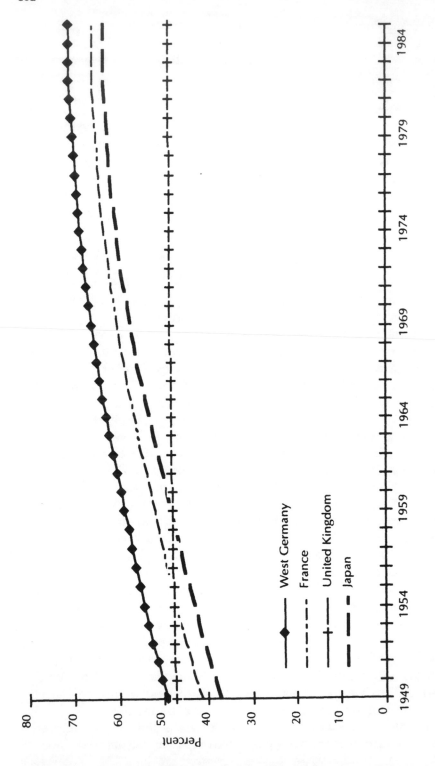

FIGURE 8.4 Productive Efficiency Relative to the United States.

or labor-augmenting (Harrod-neutral) variety. Capital-augmenting technical progress implies that the aggregate production function can be written in the form:

$$Y_t = F(A(t)K_t, L_t).$$ (15)

Thus, the benefits of technical progress are higher the higher the level of the capital stock. A country with a low level of capital stock relative to labor will not benefit as much from technical progress as a country with a high level of capital relative to labor. Capital and technical progress are complementary.

The consequence of capital-technology complementarity can be readily appreciated from our empirical results. Consider France, West Germany and Japan. They have almost the *same* estimated rate of capital augmentation of between 11 and 13 percent *per annum*. However, according to our estimates in Table 8.3, Japan has the highest average annual rate of (realized) technical progress, followed by France and then West Germany, in the same order as their respective rates of growth of capital stock. This is precisely the *complementarity* of capital and technical progress at work.

However, we should emphasize that our finding of purely capital-augmenting technical progress does not necessarily mean that the quality of labor has not improved over time or that all the investments in human capital have gone to waste. As mentioned earlier, improvements in the quality of labor may manifest themselves in the form of capital-augmenting technical progress.

We have compared the overall productive efficiencies of the Group-of-Five (G-5) countries — France, West Germany, Japan, United Kingdom, and United States — for the postwar period, using the inter-country aggregate meta-production function estimated by Boskin and Lau (1990). It is found that the United States had the highest level of overall productive efficiency for the entire period under study (1949-85). However, the productive efficiencies of France, West Germany and Japan rose rapidly from less than 50 percent of the U.S. level in 1949 to between 60 and 70 percent of the U.S. level in 1985. There is thus some evidence in favor of the hypothesis of convergence in technology.

The international differences in the levels of overall productive efficiency are not sufficient in themselves to determine international competitiveness. The latter depends on, in addition to relative productive efficiencies, relative prices of the factors of production — capital and labor — and the exchange rates. *If* the relative prices of the factors and the exchange rates have been unchanged, then clearly the United States has lost competitiveness in the postwar period. However, both the German mark and the Japanese yen have appreciated considerably, in fact, more than doubled in value, during this period, which would have offset their gains in relative productive efficiency. Moreover, the costs of capital and labor in the different countries have evolved at different rates. Further research on the relative movements of factor prices and exchange rates among the G-5 countries are required before a definite conclusion can be reached on the changes in international competitiveness.

The empirical results reported here also have significant implications for public policy. First, they reaffirm the importance of capital accumulation and technical progress as the most important source of economic growth for the Group-of-Five countries in the postwar period. This provides a strong justification for the encouragement, promotion, and support of R&D and other innovative activities, that is, for a pro-technology policy. Second, the results indicate that capital accumulation and technical progress are *complementary,* i.e., they amplify the effects of each other on productivity. This places additional emphasis on the importance of continual new gross fixed investment and argues for a pro-investment policy. Finally, given the complementarity between capital and technical progress, a pro-investment policy should go hand-in-hand with a pro-technology policy.

Notes

1. Holding the quality of the outputs and the inputs constant is easier said than done. There are, however, standard, though not necessarily perfect, ways of adjusting for the qualities of the output and the inputs.

2. In almost all such formulations, technical progress is taken to be exogenous.

3. See, for example, Abramovitz (1956), Griliches and Jorgenson (1966, 1967, 1972), and his associates Kendrick (1961, 1973), Kuznets (1971, 1973), and Solow (1957).

4. An exception is Denison (1962, 1967, 1979, 1985), who assumes that there are three inputs — capital, labor and land — and that the degree of returns to scale is 1.1, i.e., if all three inputs are increased by 1 percent, real output is increased by 1.1 percent.

5. The term "meta-production function" is due to Hayami and Ruttan (1970, 1985). See also Lau and Yotopoulos (1989) and Boskin and Lau (1990).

6. The pitfalls of maintaining the traditional assumptions of constant returns to scale, neutrality of technical progress, and profit maximization with competitive output and input markets in the measurement of technical progress and growth accounting are discussed in Boskin and Lau (1992a).

7. There are many reasons why these commodity augmentation factors are not likely to be identical across countries. Differences in climate, topography, natural resource endowment and infrastructure; differences in definitions and measurements; differences in quality; differences in the composition of outputs; and differences in the technical efficiencies of production are some examples.

8. Thus, measured inputs of any country may be converted into equivalent units of measured inputs of another country. For example, one unit of capital in country A may be equivalent to two units of capital in country B; and one unit of labor in country A may be equivalent to one-third of a unit of labor in country B. Moreover, these conversion ratios may change over time.

9. For the details of the specification of these statistical tests, see Boskin and Lau (1990).

10. It turns out that the model without first-differencing yields a Durbin-Watson statistic that is close to unity for the labor share equation, indicating serious mis-specification. The model with first-differencing yield reasonable values for the Durbin-Watson statistics. We therefore adopt the first-differenced model.

11. See, e.g., Gallant and Jorgenson (1979).

12. Readers interested in the details of these statistical tests may consult Boskin and Lau (1990).

13. However, the hypothesis of equal augmentation level parameters across countries must be interpreted carefully because differences in definitions and mea-surements, in addition to differences in the underlying qualities, will also show up as differences in the estimated augmentation level parameters.

14. The hypothesis of equal augmentation rate parameters across countries must also be interpreted carefully because they may reflect *changes* in the definitions, measurements (e.g. depreciation rates, deflators, and their errors, if any), and im-provements in the quality of complementary inputs over time, in addition to changes in the underlying quality of the inputs. Moreover, one cannot in general associate an improvement in the quality of an input with an increase in its augmentation factor. For example, an increase in the number of individuals who are computer-literate may show up as an augmentation of capital (an increase in the effective number of computers) rather than labor. Better roads may also show up as an augmentation of capital (an increase in the effective number of vehicles).

15. For a detailed discussion of how the estimate of the rate of technical progress is computed from the estimated aggregate meta-production function, see Boskin and Lau (1990).

16. The data on the capital and labor inputs are *not* adjusted for improvements in quality.

17. This section is based on Boskin and Lau (1992b).

18. The output-augmentation level parameters, $-\ell n\, A_{i0}$'s , are separately esti-mated from the data. See Boskin and Lau (1990).

References

Abramovitz, M. "Resource and Output Trends in the United States Since 1870." *American Economic Review* 46(1956): 5-23.

Arrow, K. J. "Optimal Capital Policy, the Cost of Capital, and Myopic Decision Rules." *Annals of the Institute of Statistical Mathematics* 16(1964): 21-30.

Arrow, K.J., H.B. Chenery, B.S. Minhas, and R.M. Solow. "Capital-Labor Substitution and Economic Efficiency," *Review of Economics and Statistics* 43(1961): 225-50.

Boskin, M.J. "Tax Policy and Economic Growth: Lessons from the 1980s," *Journal of Economic Perspectives* 2(1988): 71-97.

Boskin, M.J., and L.J. Lau. "Post-War Economic Growth of the Group-of-Five Coun-tries: A New Analysis," Stanford, CA: Center for Economic Policy Research, Stanford University, Technical Paper No. 217, (mimeographed), 1990.

———. "Capital Formation and Economic Growth." In *Technology and Economics: A Volume Commemorating Ralph Landau's Service to the National Academy of Engineer-ing*, 47-56. Washington, D.C.: National Academy Press, 1991.

————. "Capital, Technology, and Economic Growth." In *Technology and the Wealth of Nations*, 17-55. edited by N. Rosenberg, R. Landau and D. Mowery. Palo Alto, CA: Stanford University Press, 1992a.

————. "International and Intertemporal Comparison of Productive Efficiency: An Application of the Meta-production Function Approach to the Group-of-Five (G-5) Countries." Forthcoming in *Economic Studies Quarterly*.

Boskin, M.J., M.S. Robinson, and A.M. Hube. "Government Saving, Capital Formation and Wealth in the United States, 1947-1985." In *The Measurement of Saving, Investment, and Wealth*, 287-356, edited by R.E. Lipsey and H.S. Tice. Chicago, IL: University of Chicago Press, 1989.

Christensen, L.R., D.W. Jorgenson, and L.J. Lau. "Transcendental Logarithmic Production Frontiers." *Review of Economics and Statistics* 55(1973): 28-45.

David, P.A., and T. van de Klundert. "Biased Efficiency Growth and Capital-Labor Substitution in the U.S., 1899-1960." *American Economic Review* 55(1965): 357-394.

Denison, E.F. "United States Economic Growth." *Journal of Business* 35(1962): 109-121.

————. *Why Growth Rates Differ: Post-War Experience in Nine Western Countries*. Washington, D.C.: Brookings Institution, 1967.

————. *Accounting for Slower Economic Growth: The United States in the 1970s*. Washington, D.C.: Brookings Institution, 1979.

————. *Trends in American Economic Growth, 1929-1982*. Washington, D.C.: Brookings Institution, 1985.

Denison, E.F., and W. Chung. *How Japan's Economy Grew So Fast*. Washington, D.C.: Brookings Institution, 1976.

Denison, E.F., and R.P. Parker. "The National Income and Product Accounts of the United States: An Introduction to the Revised Estimates for 1929-80." *Survey of Current Business* 60(1980): 1-26.

Gallant, A.R., and D.W. Jorgenson. "Statistical Inference for a System of Simultaneous, Nonlinear, Implicit Equations in the Context of Instrumental Variables Estimation." *Journal of Econometrics* 113(1979): 272-302.

Goldsmith, R. *A Study of Saving in the United States*, Princeton, N.J.: Princeton University Press, 1956.

Griliches, Z., and D.W. Jorgenson. "Sources of Measured Productivity Change: Capital Input." *American Economic Review*, 56(1966): 50-61.

Hayami, Y., and V.W. Ruttan. "Agricultural Productivity Differences Among Countries." *American Economic Review* 60(1970): 895-911.

————. *Agricultural Development: An International Perspective*. rev. and exp. ed. Baltimore: Johns Hopkins University Press, 1985.

Hulten, C.R., and F.C. Wykoff. "The Measurement of Economic Depreciation." In *Depreciation, Inflation, and the Taxation of Income from Capital*, edited by C.R. Hulten. Washington, D.C.: The Urban Institute Press, 1981.

Jorgenson, D.W., and Z. Griliches. "The Explanation of Productivity Change." *Review of Economic Studies* 34(1967): 249-83.

————. "Issues in Growth Accounting: A Reply to Edward F. Denison." *Survey of Current Business* 52, Part II(1972): 65-94.

Jorgenson, D.W., F.M. Gollop, and B.M. Fraumeni. *Productivity and U.S. Economic Growth*, Cambridge, MA: Harvard University Press, 1987.

Kawagoe, T., Y. Hayami, and V.W. Ruttan, "The Intercountry Agricultural Production Function and Productivity Differences Among Countries," *Journal of Development Economics* 19(1985): 113-32.

Kendrick, J.W. *Productivity Trends in the United States*, Princeton, N.J.: Princeton University Press. 1961.

————. *Postwar Productivity Trends in the United States, 1948-1969*. New York: Columbia University Press, 1973.

Kuznets, S.S. *Economic Growth and Structure*. New York: Norton, 1965.

————. *Modern Economic Growth: Rate, Structure and Spread*. New Haven: Yale University Press, 1966.

————. *Economic Growth of Nations*. Cambridge, MA: Harvard University Press, 1971.

————. *Population, Capital and Growth*. New York: Norton, 1973.

Landau, R. "Capital Investment: Key to Competitiveness and Growth." *The Brookings Review* 9(Summer 1990): 52-56.

Lau, L.J. "On the Uniqueness of the Representation of Commodity-Augmenting Technical Change." In *Quantitative Economics and Development: Essays in Memory of Ta-Chung Liu*, 281-90, edited by L.R. Klein, M. Nerlove, and S.C. Tsiang. New York: Academic Press, 1980.

————. "Simple Measures of Local Returns to Scale and Local Rates of Technical Progress." Department of Economics, Stanford University, Working Paper, (mimeographed), 1987.

Lau, L.J. and P.A. Yotopoulos. "The Meta-production Function Approach to Technological Change in World Agriculture." *Journal of Development Economics* 31(1989): 241-69.

————. "Intercountry Differences in Agricultural Productivity: An Application of the Meta-Production Function." Working Paper, Stanford, CA: Department of Economics, Stanford University (mimeographed), 1990.

Lindbeck, A. "The Recent Slowdown of Productivity Growth." *Economic Journal* 93(1983): 13-34.

Musgrave, J.C. "Fixed Reproducible Tangible Wealth in the United States: Revised Estimates." *Survey of Current Business* 66(January 1986): 51-76.

Solow, R.M. "A Contribution to the Theory of Economic Growth," *Quarterly Journal of Economics* 70(1956): 65-94.

————. "Technical Change and the Aggregate Production Function." *Review of Economics and Statistics* 39(1957): 312-320.

9

Competitive Pressure, Productivity Growth, and Competitiveness

Nicholas G. Kalaitzandonakes, Brad Gehrke,
and Maury E. Bredahl

Introduction

Two conflicting prescriptions for policies promoting productivity growth, and so competitiveness, are being championed in the popular press. Lester Thurow (1992) in his recent book promotes industrial policy, and implicitly trade restrictions, in what he terms "An American Game Plan." An industrial policy that would create economic rents is proposed as the desired policy action to increase competitiveness. Economic rents coupled with industrial/government partnerships are seen as important contributors to increasing the ability of U.S. firms to compete in international markets.

Porter (1990) takes exactly the opposite point of view. He concludes that unfettered trade induces product innovation, productivity growth, and thus an increase in competitiveness. The reason is that creating economic rents, and by extension special-interest groups, destroys the motivation for firm actions that increase productivity growth and competitiveness. Porter even opposes such arrangement as government/industry partnerships in research and development because it might decrease the pressure for change.

Ultimately, the appropriate policy choice is an empirical question. Important issues in this investigation are the appropriate economic model and the appropriate unit of observation. As Porter concludes, "Seeking to explain 'competitiveness' at the national level, then, is to answer the wrong question. What we must understand is the determinants of productivity and the rate of productivity growth. To find answers, we must focus not on the economy as a whole but on *specific industries and industry segments*" (p. 6). The need to use industry and firm-level data to analyze competitiveness is supported by Piercy

(1982): "The problem in this present (1982) context is that the bulk of this effort has been at the macro-economic level, concerned with comparing international competitiveness between whole countries" (p. 115).[1]

The present study addresses the two points of view by asking two questions: First, does competitive pressure from intense market rivalry stimulate productivity growth and hence price competitiveness? Second, does competitive pressure stimulate productivity growth over and above what would be experienced in its absence? In other words, does competitive pressure have a structural effect on productivity growth or does it merely speed up a growth process that would take place given sufficient time?

Surprisingly, this second question has attracted little attention in the literature. Yet, it is a critical qualifying factor of the overall importance of the theme. Porter's thesis that competitive pressure fosters productivity growth appears less weighty if such influence is simply temporal. Furthermore, this issue has clear implications about the type of information necessary to address the relationship between competitive pressure and productivity growth. Cross sectional snapshots at a point in time are not sufficient to determine whether competitive pressure has long-run structural effects on productivity growth, and on price competitiveness.

The influence of competitive pressure on productivity growth is empirically investigated in this study through inter-industry comparisons of productivity growth rates for a number of Florida winter vegetable industries. Some of these industries have been subject to intense competitive pressure from imports, while others have faced little or no competitive pressure. The empirical test has a very attractive feature: Natural resources (factor endowments), technology, and other forces that often differ across nations and that affect productivity differences are not important in this analysis.

The rest of the study is organized as follows: In the second section, the decomposition of productivity growth into its structural components of technical efficiency, scale efficiency, and technological progress is illustrated. In addition, empirical evidence from previous studies that relate competitive pressure to the structural components of productivity growth is reviewed. In the third section, theoretical considerations that allow the formulation of formal hypotheses with respect to the relationship between competitive pressure and productivity growth are offered. In the fourth section, empirical evidence that addresses the formulated hypotheses is presented. In the fifth and final section, some concluding comments are offered and policy implications are discussed.

Productivity Growth and Competitive Pressure

The basic concept in the measurement of productivity growth is total factor productivity (TFP), the ratio of an index of (aggregate) output to an index of (aggregate) inputs. Any economic action that increases output while holding inputs constant is measured as an increase in TFP. Changes in TFP can be decomposed into three separate components: (a) changes in technical efficiency, (b) changes in scale, and (c) technological progress. Figure 9.1 illustrates the decomposition.

The production frontiers labeled F_1 and F_2 represent the "technically efficient" input-output combinations in the production of a given output in two periods. Point A represents the initial position of a firm in period 1 and D its final position in period 2. Since the output per unit of input employed by the firm in period 2 increased relative to that in period 1, TFP has increased between the two periods. The illustrated growth in TFP can be decomposed into the three separate adjustments: First, an improvement in technical efficiency, the movement from points A to B and second, an increase in the scale of production, which is depicted by a movement from points B to C. If this scale adjustment takes place over a production region where increasing (decreasing) returns to scale exist, TFP increases (decreases). Third, technological change results in an outward shift of the production frontier represented by a movement from point C to point D.

This decomposition can also be illustrated using cost frontiers (Figure 9.2).

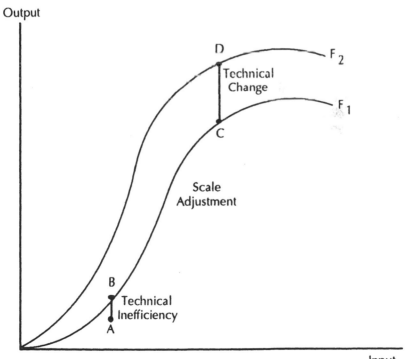

FIGURE 9.1 Decomposition of Productivity Growth Using Production Functions.

AC_1 and AC_2 are the average cost curves in periods 1 and 2 respectively. Once again, A represents the initial position of a firm in period 1 and D is the final position of the firm in period 2. Since unit cost has decreased between periods 1 and 2, TFP and price competitiveness has increased over the same time span. This increase in TFP can be decomposed into three effects as before. First, an increase in technical efficiency is depicted by a movement of the firm from point A to point B. Second, an increase in the scale efficiency is represented by a movement from point B to point C. Third, technological change shifts the cost curve downward moving the firm from point C to D. Increases in technical efficiency, improvements in scale efficiency, and technical progress, individually and collectively lead to in productivity growth as measured by TFP.

Over time, firms may utilize technically inefficient input levels, operate at inefficient scales, or use outdated technologies. It is important to understand the forces that encourage firms to reduce inefficiency, to expand and to adopt new technologies. An increasing array of empirical evidence has emerged that has found that *competitive pressure* stimulates technical efficiency, scale efficiency, and technical progress and, thus, productivity growth.

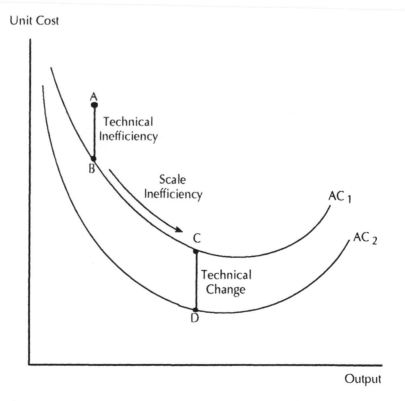

FIGURE 9.2 Decomposition of Productivity Growth Using Cost Functions.

Technical Efficiency

First, competitive pressure has been related to technical efficiency. In a series of rather influential articles Liebenstein (1966, 1973) argued that, as a general rule, firms fail to produce on their production possibilities frontier and at costs above that possible to produce a given level of output, as a result of "insufficient effort." Liebenstein further maintained that much of this inefficiency is due to a lack of competitive pressure in the market, often due to protectionism. Using firm-level data Martin and Page (1983) provided empirical support for Liebenstein's hypothesis. In particular, they related differences in the estimated efficiency levels to subsidy payments to firms in two industries in Ghana. Subsidized firms in both industries were found to exhibit substantially lower levels of technical efficiency than unsubsidized firms.

Similar corroborating empirical evidence for Liebenstein's assertions was provided by Caves and Barton (1990) who investigated the technical efficiency of more than 200 U.S. industries. They concluded that pressure from import competition raises technical efficiency in those industries where domestic production is concentrated. In separate studies, Bergsman (1974) and Balassa (1975) attempted to quantify the costs of technical inefficiencies that result from protectionist policies. Both studies concluded that limiting competition in a number of developing countries has resulted in substantial welfare costs attributable to technical inefficiencies.

Scale Efficiency

Competitive pressure has also been shown to stimulate scale efficiency. In particular, it has been proposed that competitive pressure, primarily from import competition, forces scale-inefficient firms to grow to efficient production levels or to exit the industry. Horstman and Markusen (1986) developed a two-country model where restrictive trade policies lead to insufficient import competition in a market that firms produce under increasing returns to scale and showed that the domestic industry does not rationalize, thus leading to welfare losses. Horstman and Markusen cited empirical evidence provided by Baldwin and Goreki (1983a, 1983b) to support their contention.

Baldwin and Goreki found that Canadian firms responded to the trade liberalization between 1970 and 1979 by increasing firm size or by exiting. Rationalization was particularly strong in industries that had benefited from high tariffs and where minimum efficient scales of production were large relative to those observed under restricted trade. Irvine, Sims, and Anastasopoulos (1990) used a cost function approach to estimate the gains from rationalizing production in the Canadian brewing industry following liberalization of trade between Canada's provinces. They found that such gains were, at minimum, 30 percent of existing production costs.

Further empirical evidence corroborating the hypothesis that competitive pressure has a positive influence on scale efficiency is provided by Klepper and Weiss (1992). They reported that average size changes among German manufacturing industries have been lowest in those industries that have been receiving the highest effective government assistance. Similarly, labor productivity growth rates were lower for industries that enjoyed greater protection. Klepper and Weiss concluded that "both results lend support to the foregoing hypothesis according to which increases in protection lead to inefficient scale of production, be it because of entry of new or inefficient exit of redundant firms" (p. 375).

Technological Change

Competitive pressure has been regarded as a positive influence on technological progress. Pressure from competition is the power propelling Cochrane's agricultural treadmill where firms are forced to innovate in the face of decreasing prices or exit the industry. While this hypothesis has enjoyed wide acceptability, support from firm-level empirical evidence is limited. For example, Carter and Williams (1958) in their study of investment in innovation among different firms reported that a high proportion of such investment was either in response to direct pressure from competition or "force" of example by other firms.

Summarizing, competitive pressure induces firms to improve their level of technical efficiency, to grow to a more efficient size, and to innovate to stay profitable and withstand market rivalry. Considerable empirical evidence supports this thesis. Implicit in most of these analyses is the assumption that the influence of competitive pressure is structural rather than just a mere acceleration of the growth process. Unfortunately, time-series evidence supporting this supposition is rather limited.

Theoretical Considerations

Hypothesizing that competitive pressure has a structural rather than a temporal effect on productivity growth implies that firm-level technical or scale inefficiency and laggard innovative activity may persist in the absence of pressure from competition. A natural question is why such states of inertia persist. If costs exceed the possible minimum due to wasteful use of resources, it should pay firms to reorganize production and so reduce the level of technical inefficiency. Similarly, it should pay firms to grow to sizes where average costs are minimum or to use technological innovations that would further reduce the unit costs of production. A satisfactory theory must explain why such opportunities for increased profits are not realized. Furthermore, a satisfactory theory should explain how competitive pressure stimulates technical efficiency, scale efficiency, and innovative activity when potential profits cannot do so.

To provide suitable answers to these two questions, some restrictive assumptions of the standard neoclassical model are relaxed. The assumption of perfect information is replaced by that of *imperfect and asymmetric information*. Recognizing that most technical and economic information is costly and slowly diffused provides one explanation why firms may possess different levels of information.[2] Similarly, the rigid assumption of identical producers as decision making entities is relaxed. Instead, the individuality of the firm's manager both in terms of *preferences* and *abilities*, is recognized. These alternative assumptions are not only reasonable but have also been shown to account for observed differences in efficiency and innovation among firms.[3]

The view that incomplete and asymmetric information as well as differences in preferences and capabilities among managers leads to observed differences in efficiency and innovative activity among firms allows a theoretical rationalization why competitive pressure may stimulate productivity growth when the possibility of increased profits did not. This section develops a theoretical model allowing managerial ability and preferences to differ across firms. The goal is to derive an evaluation of the persistence of less than optimum performance by each firm as an equilibrium result.

Firm Behavior

Let a particular firm be faced with the following production technology

$$Q = G(x)\, F(M,A) \tag{1}$$

where $F(\bullet)$ is assumed to be increasing at a decreasing rate in its arguments and $G(\bullet)$ satisfies the usual properties of a well-behaved production function. In this specification, x is a vector of hired labor and capital inputs, A represents managerial ability, and M denotes managerial effort. Managerial ability A is a stock concept. It is determined by an individual's personal traits and talents as well as the stock of knowledge accumulated through formal education and experience. The owner-manager allocates managerial time M among several activities, such as searching for and implementing new or improved technologies, searching for means of reducing wasteful use of inputs, exploring the possibilities of firm growth, and supervising hired labor.

The manager is assumed to actively gather information that improves the overall efficiency and productivity levels of the firm. For the specified technology, managerial ability and time are assumed to augment production in a Hicks-neutral manner. Given the latent and heterogeneous nature of managerial effort and ability, it is considered here preferable to specify M and A as shifters in the productivity of the conventional inputs x. Hence, the amount of effort M allocated by the manager and his/her existing managerial abilities A determines the levels of efficiency and productivity of the firm.[4]

The owner-manager of the firm under consideration maximizes satisfaction as a function of income π and leisure L. The income and leisure preferences of the manager are assumed to be represented by a well-behaved utility function

$$U = U(p,L). \tag{2}$$

Leisure is the residual claimer of the manager's time after managerial effort M has been accounted for. Hence, leisure is equal to $L = T^* - M$, where T^* represents the maximum time that the manager could work. Income π equals rents to management and ownership as well as any other transfer income I. The manager's budget constraint is defined by

$$p = pG(x)F(M,A) - w \cdot x + I; \tag{3}$$

where w represents input prices and p denotes the output price prevailing in the market.[5] Maximization of the utility function U in (2) subject to the budget constraint (3) results into the following first order conditions:

$$U_p = -1; \tag{4a}$$

$$U_M = lpG(x)F_M(M,A); \tag{4b}$$

$$pG_x(x)F(M,A) = -w; \tag{4c}$$

$$p - pG(x)F(M,A) + w \cdot x - I = 0; \tag{4d}$$

where $U_\lambda = \partial U / \partial \pi$, $F_M = \partial F / \partial M$, etc. For a given set of prices and transfer income, solving conditions (4a) through (4d) yields the optimal level of effort M^* that the manager of the firm selects. For firms with a particular set of income-leisure preferences and managerial abilities, the optimal level of management effort M^* may be such that firms may operate at levels below their maximum productive capacity. Thus, if the preferences and managerial abilities of the owners-managers of these firms do not change over time, such firms will exhibit *persistent* levels of inefficiency and lack of productivity growth.

Also of interest is to evaluate how the manager's effort, and hence the firm's efficiency and productivity levels, would change in response to a reduction in price p due to competitive pressure from increased market rivalry. The effect of a decrease in p on the effort M allocated by the owner-manager can be estimated through comparative static results obtained by totally differentiating (4a) through (4d). Using Crammer's rule it can be shown that

$$\partial M/\partial p = \left\{ -U_\pi GF_M \ \frac{D_{22}}{|D|} \right\} + \left\{ -U_\pi G_X F \ \frac{D_{32}}{|D|} + FG\frac{D_{42}}{|D|} \right\}; \quad (5)$$

where $|D|$ is the determinant of the matrix

$$\begin{vmatrix} U_{\pi\pi} & U_{\pi M} & 0 & 1 \\ U_{M\pi} & (U_{MM} + U_\pi pF_{MM}G) & (U_\pi pF_M G_x) & -pF_M G \\ 0 & (U_\pi pF_M G_x) & U_\pi pFG_{xx} & 0 \\ 1 & -pF_M G & 0 & 0 \end{vmatrix}$$

and D_{ij} is the cofactor of row i and column j. The first bracketed term in (5), the "substitution effect," is positive, while the second term, "income effect," is negative. Hence, the direction and the magnitude of the change in managerial effort M under a price decrease is ambiguous and depends on the relative size of the income and substitution effects. This result indicates that although a price reduction from competitive pressure may induce managers to devote a greater amount of effort to the operation of the firm, such outcome cannot be assured.

From the above it is clear that certain firms may be more productive than others as a result of their managers' abilities and efforts that are devoted to reducing technical and scale inefficiency or identifying profitable new technologies. Given stable preferences and managerial abilities, such differences in productivity among firms could persist. In the traditional neoclassical model, however, such differences are assumed to be eliminated through competitive markets and free entry.

Darwinian Process of Firm Selection

Lippman and Rumelt (1982) have shown that imperfect information is a sufficient condition for persistent differentials in profitability levels among firms under conditions of free entry and fully competitive behavior. In their theory of uncertain imitability, Lippman and Rumelt attribute insufficient entry to imperfect information and generate equilibria where stable differences in the profitability of firms and above-normal rates of return exist in the industry. Thus, imperfect information, as well as differences in preferences and capabilities among firm managers are taken here to imply varying and persistent levels of productivity among firms.

A situation where firms of varying efficiency and productivity levels coexist in the market is illustrated in Figure 9.3. Firm A is technically efficient but scale inefficient since it operates at a point of the average cost curve that unit costs are not minimum. Similarly, firm B is scale inefficient. In addition, firm B is either technically inefficient or it employs obsolete technology since it lies above the average cost frontier. In contrast, firm C is technically and scale efficient, as it

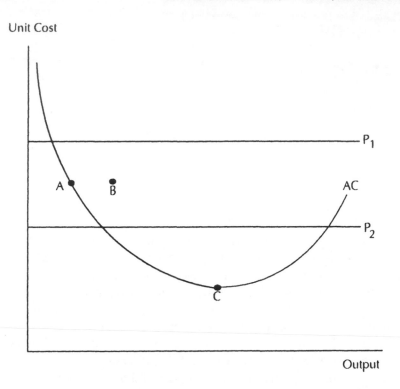

FIGURE 9.3 Darwinian process of firm survival.

operates at a minimum unit cost point, and experiences significant rents to management. Under these circumstances, firms A and B experience limited rents to ownership and management at price p_1.

Under stationary preferences and economic conditions, the situation illustrated above can be rather stable. A changing economic environment that increases the competition in the market, however, can activate an adjustment process. Increased rivalry among firms can be initiated either by decreasing demand conditions or entry of competitors with cost advantage over existing firms, such as low-cost import competition. In either case, the end result is a reduction in the market price p. Under the new market price p_2, firms A and B experience less than competitive returns to their productive factors and, hence, they cannot maintain their existing position in the long run. The strategies available to the firms are different depending on their initial position. Firm A can elect to increase its size and hence reduce its scale inefficiency by taking advantage of existing scale economies or exit the industry. Firm B can elect to reduce its technical inefficiency, innovate, increase its scale efficiency, or choose a combination of the above strategies that would, at minimum, reduce unit costs to a level equal to p_2. Alternatively, firm B may choose to exit the industry if the preferences of the owner-manager are such that additional effort under the changed economic conditions does not maximize his/her utility.

Whether firms A and B choose to become more technically and scale efficient, innovate, or exit the industry, however, leads to *aggregate* productivity growth. In the case where firms A and B choose to remain in the industry by increasing their average productivity, the industry's output per unit of (aggregate) input employed increases, thus manifesting aggregate productivity growth. Similarly, in the case where firms A and B elect to exit the industry, the industry's output per unit of input rises as the firms with the least efficient input use cease production. Thus, once again aggregate productivity growth is experienced in the industry.

Two main conclusions can be drawn from the above theoretical considerations. First, the influence of competitive pressure on individual firms is ambiguous and it is determined by the preferences and managerial abilities of the owner-manager of each firm. Hence, *firm-level* information may or may not support the hypothesis that competitive pressure stimulates productivity growth. Second, the influence of competitive pressure on the industry's productivity growth is unequivocal. Competitive pressure through a Darwinian selection process leads to increased technical efficiency, scale efficiency, and innovation of some firms and exit of others. Under such conditions, the industry experiences productivity growth. Hence, *industry-level* information is appropriate in testing the influence of competitive pressure on productivity growth since the direction of the hypothesized relationship between competitive pressure and productivity growth at the industry level is unequivocal.

In conclusion, the following testable hypothesis can be formulated: Competitive pressure has a positive influence on an industry's productivity growth. In its absence, inefficiencies and laggard innovative activities can persist even in the presence of potential for increased profit. Thus, competitive pressure may indeed have a structural influence on productivity growth not replaceable by the passage of sufficient time.

Empirical Analysis and Results

As previously emphasized, in order to test the hypothesis that competitive pressure has a structural effect on productivity growth sufficient time series industry-level information must be employed. In this context, Florida fresh winter vegetable industries offer an excellent opportunity to examine the relationship between competitive pressure and productivity growth.

Over the 1970-85 period under consideration, production costs were similar among U.S. domestic producers of fresh winter vegetables while Mexican producers enjoyed an absolute competitive advantage in production costs (Simmons et al., 1976; Zepp and Simmons, 1979; Buckley et al., 1986). For those Florida crops facing only domestic competition, market boundaries were mainly delineated by transportation costs, crop perishability, and production timing differentials. In contrast, production-cost advantages enabled vegetables imported from Mexico to compete in markets traditionally supplied by Florida, such as the north and the northeast regions of the United States. The intensity of the competition between Florida and Mexico is well documented in Bredahl et al., 1983 and Buckley et al., 1986.

Given these differential patterns of competition, vegetable crops produced in Florida can be partitioned into two independent sets of crops based on the extent of competitive pressure. Cucumbers, peppers, squash, and tomatoes faced direct competition from Mexican imports and so experienced considerable competitive pressure from one set. Cabbage, celery, sweet corn, eggplant, leaf crops, potatoes, radishes, and watermelons faced virtually no foreign competition and only limited domestic competition from a second set.

A comparative analysis of productivity growth rates over the sixteen year period 1970-85 and across the two groups of crops provides a direct test of the relationship between competitive pressure and productivity growth. In addition to the natural division of Florida winter vegetable crops into two mutually independent groups with differential levels of competitive pressure, two added factors strengthen the comparative analysis of productivity growth. First, the quality of the factor endowments across the two groups of vegetables is homogeneous. Second, as documented in *Florida Agriculture in the 80's: Vegetable Crops*, similar new technologies and improved cultural practices were available for most crops during the period of the analysis. Such new technologies included improved cultivars, utilization of plastic mulch, high density plantings, and new irrigation and pest control practices. The similarities in the nature of the available technologies further suggest that no major differences existed in the size of the initial investment requirements and the risks associated with their adoption. Hence, the availability of improved technologies, the size of initial investment, and the risks of adopting new technologies are not expected to have influenced the productivity rates across the two sets of Florida vegetable crops. Given these conditions, it is possible to isolate, to a large extent, the relationship between competitive pressure and productivity growth.

Empirical Model

Productivity growth for the two sets of Florida winter vegetable crops was measured using the index numbers approach to TFP measurement. For a single-output production process, TFP is the ratio of output Q to an index of aggregated inputs X. Hence, the proportionate rate of growth of TFP should be equal to the difference between the proportionate rates of growth in Q and X. Along these lines, the Divisia index of TFP growth is defined as

$$T\dot{F}P = \dot{Q} - \dot{X} \tag{6}$$

$$\dot{X} = \sum_i S_i \dot{x}_i, \tag{7}$$

S_i is the budget share of the ith input of production, and a dot over a variable denotes proportional rate of growth. Therefore, the Divisia index of TFP growth is the residual growth in output not accounted for by the growth in inputs.

Calculation of TFP growth through (6) and (7) presupposes continuous time series data that, in practice, are not available. For that reason, the above continuous expressions are generally approximated using discrete data. Tornqvist-Theil indices are used here as discrete approximations to (6) and (7) and are specified as

$$\ln(TFPT/TFP_0) = \ln(Q_T/Q_0) - \ln(X_T/X_0) \text{ and} \tag{8}$$

$$\ln(XT/X_0) = \frac{1}{2}\sum_i (S_{iT} + S_{i0}) \ln(x_{iT}/x_{i0}); \tag{9}$$

where S_{it} is the cost share of input i in period t. Equations (8) and (9) provide the basis for estimating TFP growth for all Florida winter vegetable industries under consideration.

Computation of TFP indices for each crop required data on output, costs, and input quantities over the 1969-70 to 1984-85 period. Output was measured as yield per acre for each vegetable crop. Production costs were obtained from Brooke (1969-79), Taylor (1982), Taylor and Wilkowske (1983), and Taylor and Wilkowske (1984). Input categories used in the computation of TFP included seed, fertilizer, agricultural chemicals, labor, energy, capital services, and a miscellaneous inputs category. Implicit input quantity indices were generated using production expenditure data and price indexes reported in *Agricultural Prices*.

TFP indices were calculated according to equation (8) for each crop over the 1970 to 1985 period, and are available from the senior author.[6] In order to be able to compare the productivity growth experienced by the vegetable industries in the two sets under consideration, average rates of productivity growth were necessary. Zohar and Luski (1987) have outlined several different ways in which average productivity growth rates may be calculated. Suggested measures include the use of regression, the arithmetic average, the geometric average, and the geometric average of the beginning and ending periods of the annual TFP indices.

In the present analysis, the use of regression was preferred since it is the only method that allows average annual rates of TFP growth to be derived while accounting for exogenous factors that may cloud the calculation of TFP indices. Adverse weather is one such factor that can complicate the computation of TFP since it can cause variations in output that are unrelated to input usage. Within this context, average annual rates of productivity growth were derived in this study through regression analyses. In particular, the logarithmic expression of the calculated TFP series for each crop was regressed against a trend variable and a weather dummy variable. The weather dummy is incorporated in the regression in order to capture the effects of abnormal weather on TFP. The relationship between unreasonably low or high yields and weather is documented through annual issues of *Vegetable Summary* where significant weather variations and their effects on annual yields are reported.

Empirical Results

The estimated average annual rates of productivity growth for all the crops for the period 1970-85 are presented in Table 9.1. The average annual rates of productivity growth for all the Florida fresh vegetable crops that faced Mexican competition, and hence significant competitive pressure, are substantial: Annual rates of TFP growth for these crops varied from a minimum of 1.6 percent to a maximum of 6.5 percent. In contrast, few of the vegetable crops that faced limited competitive pressure exhibited noticeable productivity growth. Of the

TABLE 9.1 Average Annual Rates of Productivity Growth for Florida Vegetable Crops, 1969/70 to 1985/86.

Crop	Region	Total Factor Productivity Growth (%)
Limited Competitive Pressure		
Cabbage	Hastings	0.55
Celery	Central Florida	0.74
	Everglades	0.18
Sweet Corn	Central Florida	0.12
	Everglades	0.41
	Lower East Coast	1.85
Eggplant	Palm Beach	0.73
Leaf Crops	Central Florida	2.94
	Everglades	-1.85
Potatoes	Dade County	-0.76
	Hastings	2.91
Radishes	Everglades	-2.06
Watermelon	Imm/Lee	-2.96
Significant Competitive Pressure		
Cucumbers	Imm/Lee	6.50
Bell Peppers	Imm/Lee	2.18
	Palm Beach	2.82
Squash	Dade County	1.56
	Imm/Lee	2.10
	Palm Beach	4.25
Tomatoes	Dade County	2.75
	Imm/Lee	5.25
	Manatee/Ruskin	4.20

13 crops that did not face competitive pressure, 10 exhibited very small positive or negative rates of total factor productivity growth. The negative rates of productivity growth could be explained by the development of disease and pest infestations that required an increase in control measures (inputs) without a corresponding increase in output, as well as the exhausting use of marginal land, as in the case of watermelon production. The exceptions that exhibited measurable increases in total factor productivity are leaf crops grown in central Florida, potatoes produced in Hastings, and sweet corn grown in the lower-east coast area; but, all exhibited annual rates of growth of less than 3 percent.

Comparing the rates of productivity growth between the two sets of vegetable crops provides further evidence of the effect of competitive pressure on total factor productivity growth. The average rate of growth for the vegetable industries that faced little competitive pressure equaled only 0.2 percent per year. In contrast, the average growth in productivity for all the vegetable industries that faced substantial competition in their market equaled 3.5 percent per year. The differences in productivity growth experienced by the two groups of industries are substantial. It is important to remember these differences were experienced despite the similarities in factor endowments and technology availability.

Furthermore, the substantial differences in productivity growth rates remained large over an extended period of time with no obvious tendencies of converging to similar levels. Thus, the empirical evidence provided by the Florida fresh winter vegetable industries supports the stated hypothesis that competitive pressure has a positive *structural* influence on productivity growth.

The empirical results are limited, however, in that assessment of the individual contribution of technical efficiency, scale efficiency, and technical progress to the measured TFP growth of Florida vegetable crops is not possible. Panel data are necessary for the decomposition of TFP into its structural components. Such information, however, is not available for the Florida vegetable industry.[7]

Conclusions and Implications

This paper contributes to understanding the process of firm and industry adjustment to competitive pressure that might arise from import competition or from a reduction in policy-determined prices. Several previous studies have argued that individually competitive pressure stimulates technical efficiency, scale efficiency, and/or technological progress. As developed in this study, improvements in technical and scale efficiency as well as technological progress collectively increase productivity. Therefore, competitive pressure has been indirectly considered a stimulus to productivity growth. An important question addressed in the paper is that of the structural effect of competitive pressure on productivity growth. If productivity growth simply occurs more slowly in the absence of competitive pressure, then much of the economic gains thought to arise from competitive pressure are likely overstated.

Imperfect and/or asymmetric information, as well as differences in the abilities and preferences of the firms' managers, allows stable equilibria to exist with firms of varying levels of productivity coexisting in a given industry. Imperfect information at the industry level explains the failure of firms to enter the market and, so, to drive the industry to the equilibrium predicted by the perfectly competitive model. Under these conditions, persistent lack of productivity growth is possible. Thus, competitive pressure, as a force that induces change, could indeed have a structural effect on productivity growth.

It was further shown that at the firm-level the direction of the response to competitive pressure, as indicated by a price decline, cannot be unambiguously determined. At the industry level, however, the result of an increase in competitive pressure must be an unambiguous increase in total factor productivity. The empirical results supported the conclusion that competitive pressure induces productivity growth and also that it has a structural impact. In the absence of competitive pressure, lower rates of productivity growth are noted, and those lower rates persisted over a long time period.

The theoretical model and the empirical results raise a number of questions about analysis of reactions of firms, and industries, to competitive pressure. The theoretical and empirical results suggest, for example, that much of the research on trade liberalization may have understated the gains. If significant gains from rationalization of the industry follow trade liberalization, the gains to increasing competitive pressure could be quite large. The flipside of that question, is that the cost of protectionist trade policies are much larger than have been estimated.

Notes

1. Competitiveness in the context of these comments, and of this study, is price competitiveness. Price competitiveness for an industry is created through reduction in factor costs, improvements in factor productivity, and variations in exchange rates. For a more complete discussion on the notion of price competitiveness see Francis.

2. Imperfect or incomplete information has been explicitly related to technical inefficiency (Muller, 1974, Shapiro and Muller, 1977), scale inefficiency (Stefanou, 1989), and laggard innovation (Jensen, 1982, Hierbert, 1974, and Feder and Slade, 1984).

3. Managerial abilities have been associated to observed differences among firms in terms of technical efficiencies (Kalaitzandonakes and Moore, 1992) as well as innovative activities (Kislev and Shchori-Bachrach, 1973). Similarly, individual preferences for acceptable risk, income and leisure have been explicitly used to explain differences in technical efficiencies (Liebenstein, 1973, Martin and Page, 1983) and scale inefficiencies (Balassa, 1975) that are empirically observed among firms.

4. Firms whose managers devote less effort to the operation relative to others with equal abilities will be less efficient and productive within the proposed model. The welfare implications of allocating less managerial effort are not addressed in this study. In order to deal with the issue of whether inefficiency induced by less effort leads to welfare loss, the distinction between private and social costs must be made. For a discussion on this issue see Stigler (1976).

5. For a formulation that allows for external markets of managerial services see Martin and Page (1983).

6. These derivations extend previous results of Kalaitzandonakes and Taylor (1990).

7. For an illustration of TFP decomposition in its structural components see Nishimizou and Page (1982).

References

Balassa, B. "Trade, Protection, and Domestic Production: A Comment." In *International Trade and Finance: Frontiers for Research*, edited by P.B. Kenen. Cambridge: Cambridge University Press, 1975.

Baldwin, J.R., and P. Goreki. "Entry and Exit to the Canadian Manufacturing Sector: 1970-79." Economic Council of Canada, Discussion Paper No. 225, 1983a.

———. "Trade, Tariffs, Entry and Relative Plant Scale in Canadian Manufacturing Industries: 1970-79." Economic Council of Canada, Discussion Paper No. 232, 1983b.

Bergsman, J. "Commercial Policy, Allocative Efficiency, and X-efficiency." *Quarterly Journal of Economics* 88(1974): 409-33.

Bredahl, M.E., J.S. Hilman, R.A. Rothenberg, and N. Gutteriez. "Technical Change, Protectionism and Market Structure: The Case of International Trade in Fresh Winter Vegetables." University of Arizona, Technical Bulletin No. 249, 1983.

Brooke, D.L. "Costs and Returns from Vegetable Crops in Florida with Comparisons." Food and Resource Economics Department, Economic Information Report, Gainesville: University of Florida, 1969-79.

Buckley, K.C., J.J. VanSickle, M.E. Bredahl, E. Belibasis, and N. Guteriez. *Florida and Mexico Competition for the Winter Fresh Vegetable Market*. USDA Economic Research Service. Agricultural Economic Report No. 556, 1986.

Carter, C.F., and B.R. Williams. *Investment in Innovations* London: Oxford University Press, 1958.

Caves, R.E., and D.R. Barton. *Efficiency in U.S. Manufacturing Industries*. Cambridge, MA: The MIT Press, 1990.

Feder, G. and R. Slade. "The Acquisition of Information and the Adoption of New Technology." *American Journal of Agricultural Economics* 66(1984): 312-20.

Hiebert, I.D. "Risk, Learning, and the Adoption of Fertilizer Responsive Seed Varieties." *American Journal of Agricultural Economics* 56(1974): 764-68.

Horstman, I.J., and J.R. Markusen. "Up the Average Cost Curve: Inefficient Entry and the New Protectionism." *Journal of International Economics* 20(1986): 225-47.

Institute of Food and Agricultural Sciences. *Florida Agriculture in the 80's: Vegetable Crops*. Gainesville: University of Florida, 1983.

Irvine, I.J., W.A. Sims, and A. Anastasopoulos. "Interprovincial Versus International Free Trade: The Brewing Industry." *Canadian Journal of Economics* 23(1990): 332-47.

Jensen, R. "Adoption and Diffusion of an Innovation of Uncertain Profitability." *Journal of Economic Theory* 27(1982): 182-93.

Kalaitzandonakes, N.G., and K. Moore. "Managerial Ability and Technical Efficiency." University of Missouri, (mimeographed), 1992.

Kalaitzandonakes, N.G., and T.G. Taylor. "Competitive Pressure and Productivity Growth: The Case of the Florida Vegetable Industry." *Southern Journal of Agricultural Economics* 22(1990): 13-21.

Klepper, G., and F.D. Weiss. "Protection and International Competitiveness: A View from West Germany." In *International Productivity and Competitiveness*, edited by B. Hickman. New York: Oxford University Press, 1992.

Kislev, Y., and N. Shchori-Bachrach. "The Process Innovation Cycle." *American Journal of Agricultural Economics* 55(1973): 28-37, 1973.

Liebenstein, H. "Allocative Efficiency vs X-efficiency." *American Economic Review* 56(1966): 392-415.

———. "Competition and X-efficiency: Reply." *Journal of Political Economy* 81(1973): 765-67.

Lippman, S.A., and R.P. Rumelt. "Uncertain Imitability: An Analysis of Interfirm Differences in Efficiency under Competition." *Bell Journal of Economics* 13(1982): 418-38.

Martin, J.R., and J.M. Page. "The Impact of Subsidies on X-efficiency in LDC Industry: Theory and an Empirical Test." *Review of Economics and Statistics* 65(1983): 608-17.

Muller, J. "On Sources of Measured Technical Efficiency: The Impact of Information." *American Journal of Agricultural Economics* 56(1974): 730-38.

Nishimizou, M., and J.M. Page. "Total Factor Productivity Growth, Technological Progress, and Technical Efficiency Change: Dimensions of Productivity Change in Yugoslavia, 1965-78." *Economic Journal* 92(1982): 920-36.

Piercy, N. *Export Strategy: Markets and Competition.* London: University of Wales Institute of Science and Technology, George Allen & Unwin, 1982.

Porter, M. *Competitive Advantage of Nations.* New York: The Free Press, 1990.

Shapiro, K., and J. Muller. "Sources of Technical Efficiency: The Roles of Modernization and Information." *Economic Development and Cultural Change* 25(1977): 293-310.

Simmons, R.L., J.L. Pearson, and E.B. Smith. *Mexican Competition in the U.S. Fresh Winter Vegetable Market.* USDA Economic Research Service Agricultural Economic Report No. 348, 1976.

Stefanou, S. "Learning, Experience, and Firm Size." *Journal of Economics and Business* 41(1989): 283-96.

Stigler, G.J. "The Xistence of X-efficiency." *American Economic Review* 66(1976): 213-16.

Taylor, T.G. "Costs and Returns from Florida Vegetable Crops, Season 1980-81, with Comparisons." Food and Resource Economics Department, Economic Information Report No. 159, Gainesville: University of Florida, 1982.

Taylor, T.G. and G.H. Wilkowske. "Costs and Returns from Florida Vegetable Crops, Season 1981-82, with Comparisons." Gainesville: Food and Resource Economics Department, Economic Information Report 161, University of Florida, 1983.

———. "Productivity Growth in the Florida Fresh Winter Vegetable Industry." *Southern Journal of Agricultural Economics* 16(1984): 55-61.

Thurow, L. *Head to Head*, New York: William Morrow and Company, 1992.

U.S. Department of Agriculture. *Vegetable Summary* Orlando, Florida: Agricultural Marketing and Crop Reporting Service, 1962-85.

U.S. Department of Agriculture, *Agricultural Prices*. Washington D.C.: Crop Reporting Service and Economics and Statistics Service. 1969-1985.

Zepp, G.A., and R.L. Simmons. "Producing Fresh Winter Vegetables in Florida and Mexico." USDA Economics, Statistics and Cooperatives Service Report No. 72, 1979.

Zohar, U., and I. Luski. "A Note on the Measurement of the Slowdown in Total Factor Productivity." *Applied Economics* 19(1987): 1211-19.

PART FOUR

Assessments of the Competitiveness of National Food Sectors

10

Assessing the International Competitiveness of the United States Food Sector

Stephen MacDonald and John E. Lee, Jr.

Trying to determine the competitiveness of an entire industry as complex as the United States food sector is an ambitious undertaking. The complexity of the food sector and world food trade, given the distortions created by food and trade policies makes such analysis difficult. Our approach is simply to examine and interpret some of the most readily available productivity and trade data. It is difficult to definitively judge the competitiveness of the U.S. food sector. Given the data easily accessible, we conclude that the U.S.'s overall comparative advantage in food exports is likely greater than its actual competitive position in world markets because of barriers to trade and policy-related market distortions. U.S. farm productivity is high and growing, but the U.S. competitive position has been slipping in semiprocessed food and agricultural products.

What Does Competitiveness of the U.S. Food Sector Mean?

Some of the difficulties in addressing the question of competitiveness are obvious. First of all, the sector has many parts: input suppliers, producers, processors, wholesalers, retailers and others. Within each of these general parts are many different products and many different firms producing those products. There are competitive and noncompetitive firms, and competitive and noncompetitive locations within a country, especially in a large country such as the United States. When we think of the large numbers of products, levels of processing, and the diversity of firms, it becomes difficult to generalize about the competitiveness of "the food sector."

Second, when we speak of competitiveness, are we talking about real comparative advantage or competitiveness in the context of the existing distortions that mask underlying comparative advantage? Unless one can estimate and adjust for the net effects of existing distortions, does "competitiveness" have any meaning, and can we say anything meaningful about comparative advantage?

Third, when we refer to the competitiveness of the food sector, are we talking about competitiveness of products or competitiveness of firms? U.S. food firms play a big role in world food markets, but most of this role is in foreign investment, foreign processing plants, partnerships, franchises, and licenses, rather than in exports of food products. If the firms in a U.S. food processing sector dominate foreign markets through a variety of forms of business and investment, but do little importing or exporting, what does that tell us about competitiveness of that sector?

To complicate the question, it is increasingly difficult to distinguish the nationality of a firm. In fact, does the nationality of a firm mean anything? For addressing competitiveness, does it mean anything that Pillsbury, a so-called "U.S. American" firm, is bought out by Grand Met, a "British" firm? If we say that U.S. food manufacturing firms are competitive because they do a large amount of foreign (nontrade) business, what changes if the majority of stock is bought by Europeans or Saudis, or if the company decides to move its corporate headquarters to the Bahamas?

Take the case of automobiles. If the Japanese find that they can compete more effectively by building their cars in the United States, both for domestic sales and export, couldn't we say that U.S. automobile manufacturing is competitive, even if old line U.S. companies are still declining? What changes if GM buys majority stock in Toyota, Toyota buys controlling interest in GM, or GM contracts with Japanese management to run GM?

The point of all this is that one must look at where products are produced or manufactured, where they flow and why, to investigate the issue of competitiveness. We have to assume that firms in any country, whether locally owned or foreign-owned, behave rationally; that is, they obtain their inputs from the lowest-cost source, given the set of constraints within which they operate. What firms do in terms of location, entrepreneurial initiatives, and foreign investment are all important topics, but they confuse the issue of competitiveness of the U.S. food production and processing sector. To study competitiveness, one must look at product flows (imports and exports) and the reasons for those flows.

Other Conceptual Considerations

Theory would suggest that, unless there are some offsetting factors, processing reduces bulk or weight to be transported. In other words, transportation is a smaller proportion of the product value for high-value and value-added products than for bulk raw products.

When one observes the data in this regard, there is a mixed story. There is a tendency for industrial nations to import bulk raw materials from developing nations; however, there are some exceptions. The United States ships tree logs from the West Coast to Japan. Japan processes the logs into a variety of wooden containers and products and ships the high-value, light-weight finished products back to the United States. The explanation must be that the logs are so cheap that, even when transportation costs are added, the raw materials are such a small part of the total cost of the final product, that comparative advantage in producing the final product more than offsets the costs of wood. In other cases, the perishable nature of the final product (bread, for example) could explain why trade tends to be primarily raw or intermediate-processed products, although one could make the opposite case for dairy products.

A second conceptual consideration involves the ongoing debate about whether trade in value-added or further-processed products is "better" than trade in raw or unprocessed products. The standard argument is that emphasis (even subsidies) should be put on value-added exports because of the additional employment and GNP generated by the activities that add value. Dressed poultry production generates almost nine times as much personal income and employment as an equal value of raw corn, while flour generates almost 2.5 times as much personal income and employment as an equal value of raw wheat (Schulter and Edmondson, 1989). Such data are commonly used by grain millers, bakery unions, and others to support their case for increased value-added exports. Some policymakers are seduced by these arguments and put pressure on program agencies to do more to promote value-added exports.

Value-added exports are beneficial to a country if the country has a comparative advantage in value-adding activities. But if a country has a comparative advantage in producing wheat, corn, or other raw bulk products, and does not have a comparative advantage in further processing, forcing or subsidizing exports of further-processed products would represent an inefficient use of resources. In other words, whether a nation should export a given product in bulk or value-added form depends on the comparative advantage or disadvantage that nation has in the production of that raw commodity and in the successive stages of processing and services that add value. Thus, whether the United States should export wheat, flour, or baked goods should depend on comparative advantage in wheat production, comparative advantage in flour milling, and comparative advantage in producing bakery products. The difficulty in determining actual comparative advantages of nations in various products, bulk or value-added, from observed trade data is that the mix of products actually traded is driven by a combination of underlying comparative advantages and distortions introduced by government interventions.

Measuring Competitiveness

There are two approaches one could take to empirical examination of the competitiveness of the U.S. food sector. One approach is to examine the levels and trends of total factor productivity in the U.S. food sector relative to levels and trends in productivity of other sectors of the U.S. economy and of food and nonfood sectors in other economies. If the data were adequate to the task, this approach would reveal the underlying comparative advantage of various U.S. food products in world markets. Unfortunately the data are not very complete. The ideal data set would include levels and rates of change of total factor productivity for components of the U.S. food sector and the rest of the U.S. economy, and the food sector and other sectors of other countries. Good measures of total factor productivity are available for the United States, Brazil, and EC agricultural sectors, some estimates of multifactor productivity for other sectors of the U.S. economy, and estimates of labor productivity of food processing sectors of OECD countries.

Arnade (1992) calculated multifactor productivity indices for Brazil for 1968 to 1987 under two assumptions: constant returns to scale and non-constant returns to scale. In Table 10.1 the indices are shown using constant returns to scale so they can be compared to U.S. productivity measures. The indices are normalized to equal 100 in 1968. Therefore, the indices in Table 10.1 compare productivity growth in both countries relative to their respective base, but the levels are not compared. While it is likely that the United States started the period with a higher absolute level of productivity, the numbers are not available. For 1987, the U.S. index is 148 and the Brazilian index is 163, suggesting that Brazilian productivity grew faster over the twenty-year period, especially from 1982 onward. Since total crop and livestock productivity grew faster than crop productivity in Brazil, one must assume that livestock productivity grew even faster than the aggregate.

Indices calculated by Bureau suggest that while farm productivity is nearly one-third higher in the United States than in the EC-10, the rate of growth in productivity from the mid-1970s to the late 1980s has been about the same, at 2.2 percent per year. In some countries of the EC-10 (France, Italy, Denmark, and Greece), productivity grew faster than in the United States. What these numbers suggest is that, while U.S. agricultural productivity may be high in absolute levels, its growth rate is below Brazil's and the EC-10's, suggesting that the U.S.'s comparative advantage is eroding slightly relative to those countries.

For the food processing industry, 1985 estimates for labor productivity show that among OECD countries, only Belgium, Ireland, the Netherlands, and the United Kingdom increased their productivity more than the United States in the first half of the 1980s (Table 10.2). Handy and Henderson (1991) have used data that show that while there are great similarities between the U.S. and EC food manufacturing sectors, labor productivity appears to be higher in the

TABLE 10.1 Productivity Indices (constant returns to scale), Brazil (1968=100)

Year	Crops and Livestock			Crops			U.S. Productivity[3]
	Outputs[1]	Inputs[2]	Productivity	Outputs	Inputs	Productivity	
1968	100	100	100	100	100	100	100
1969	103	97	106	103	101	102	101
1970	109	98	111	109	102	107	99
1971	111	101	111	114	106	107	108
1972	117	106	111	117	113	104	110
1973	120	108	112	120	118	102	117
1974	132	113	117	136	128	106	105
1975	135	119	113	136	139	102	113
1976	138	124	112	138	139	99	110
1977	149	128	117	151	140	108	121
1978	141	129	109	138	139	99	113
1979	148	130	114	148	139	106	118
1980	163	141	115	163	159	103	113
1981	166	140	119	162	149	105	129
1982	176	132	133	171	143	120	130
1983	178	143	124	179	146	122	110
1984	185	138	135	192	150	128	134
1985	205	133	155	216	150	144	144
1986	187	136	138	195	147	132	143
1987	209	128	163	216	144	150	146

[1]Outputs index represents a weighted average of outputs with revenue shares as weights.
[2]Inputs index represents a weighted average of inputs with cost shares as weights.
[3]Based on Economic Research Service productivity indices for U.S. agriculture. Indices assume constant returns to scale and are normalized so that 1968=100.
Source: Arnade (1992).

United States, possibly giving the United States a potential comparative advantage. Sales per employee in food manufacturing averaged $218,000 for the United States (1987) and $136,000 for the EC (1986). For the same years, value added per employee was $84,000 in the United States and $35,000 in the EC. While the value added per employee varied widely from one type of food processing to another, it was consistently higher for the United States than for EC processors. EC integration will permit substantial improvement in these measures as food manufacturers capture scale economies and restructure their industry to serve a larger market.

Productivity estimates for U.S. industry prepared by the Bureau of Labor

TABLE 10.2 Productivity in Food Processing Industries in Selected OECD Countries, 1985 (1980=100)

Country	Production	Employment	Labor Productivity
Australia	100	90	111
Austria	106	90	118
Belgium	121	92	132
Canada	105	89	117
Denmark	120	110	110
Finland	111	101	110
France	107	105	102
Germany	109	92	118
Ireland	119	81	146
Italy	103	95	108
Japan	104	102	102
The Netherlands	110	88	125
Norway	64	100	65
Sweden	104	97	106
United Kingdom	104	80	130
United States	117	96	122

Source: OECD.

Statistics (BLS) suggest that single-factor productivity measures, such as labor productivity, should be interpreted cautiously as indicators of competitiveness. For example, BLS data show high growth rates in labor productivity for Food and Kindred Products Manufacturing in the United States, but low growth rates in the productivity of other factors. As a result, multifactor productivity in U.S. food manufacturing grew only 0.6 percent per year from 1949 to 1988 and only 0.7 percent annually from 1979 to 1988. Thus, it appears that productivity in U.S. food manufacturing has grown less rapidly than in agriculture. Without comparable data for other countries, it is hard to draw conclusions about the competitiveness of the U.S. food manufacturing industry in world markets.

Bureau calculated single-factor productivity indices for intermediate inputs, land, capital, and labor for EC-10 agriculture and compared them with the United States. While the United States far outstripped the EC-10 in labor productivity, it just edged the EC-10 in productivity of intermediate inputs and fell well below EC-10 productivity of land and capital. This reflects, in part, the relative scarcity of production factors in the two cases. If this relationship holds for food manufacturing in the EC-10, then total factor productivity of food manufacturing in the EC could be more comparable to that for the United States than comparisons of labor productivity alone would suggest.

Trends in Agricultural Trade

A second approach to examining the competitiveness of the U.S. food industry is to observe and interpret trade data. This analysis groups products by value and processing level. By examining product trade with varying degrees of value added, one can observe how the United States is faring in trade absolutely and relative to the rest of the world. One should expect some increase in higher-valued exports over time as real incomes of nations rise. This is because as consumer's demands for basic foods are satisfied, further growth in food expenditures, especially in more developed countries, tends to favor a greater mix of high-valued, highly processed foods.

For this analysis, agricultural products will be divided into low-valued products and high-valued products. Low-valued products include bulk commodities, such as grains and oilseeds, and raw materials, such as cotton and tobacco. High-valued products generally include three groups: high-value unprocessed foods (e.g., eggs, fresh fruits and nuts, and fresh vegetables); semiprocessed products (e.g., flour, oilseed products, and meats); and highly processed products (e.g., prepared and preserved meats, dairy products, bakery products and prepared foods). Most of the data are from an Economic Research Service database containing primarily data from the FAO and UN.

World trade in agricultural products totalled $312 billion in 1990. Highly processed products accounted for $105 billion, high-valued unprocessed accounted for $80 billion, semiprocessed accounted for $71 billion and low-valued bulk accounted for $56 billion. While much media attention is given to trade and trade negotiations dealing with raw, bulk commodities, world trade in high-value food products is much more important.

Comparative Content of World
and U.S. Agricultural Trade

Global trade in agricultural products grew steadily in the 1970s, flattened out in the early 1980s, and resumed growth after 1985 (Figure 10.1). The share of high-value exports grew slightly at the expense of raw unprocessed bulk products, especially after 1985. Bulk unprocessed products have dominated U.S. agricultural exports (Figure 10.2), especially in the 1970s; exports of raw agricultural products surged, while exports of high-value agricultural products grew steadily but slowly. Thus, the spread between high-value and total exports was greatest in the late 1970s and very early 1980s. During the 1980s, U.S. exports of bulk grains and raw products dropped substantially before stabilizing after 1986. However, U.S. exports of high-value agricultural exports declined modestly in the early 1980s, then stabilized, and resumed growth in the late 1980s. In 1992, high-value U.S. agricultural exports exceeded the value of bulk unprocessed exports for the first time.

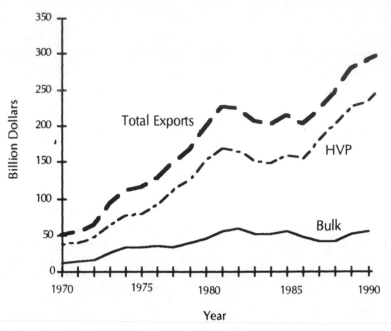

FIGURE 10.1 World Agricultural Exports, 1970-90

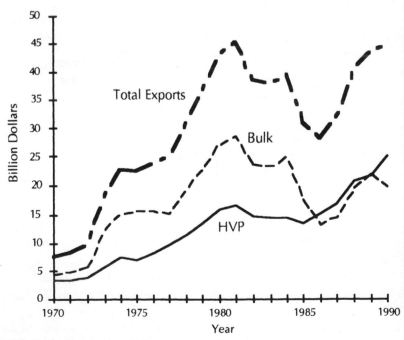

FIGURE 10.2 U.S. Agricultural Exports, 1970-1990

The United States still lags overall world trade in terms of high-value and value-added content. Approximately 45 percent of U.S. agricultural exports are bulk versus 20 percent for the rest of the world; 23 percent of U.S. agricultural exports are highly processed versus 32 percent for the rest of the world. Nevertheless, the more recent mix of products exported stands in sharp contrast to the situation in the mid-1970s to the mid-1980s. This change over the past decade may suggest growing competitiveness of high-value U.S. exports, at least relative to bulk products. High-value exports from the United States have been aided by rising incomes in Japan, some other East Asian countries, and, more recently, Mexico. While the EC is still the largest market for U.S. high-value agricultural exports, its share is declining.

Comparative Growth Rates of World and U.S. Agricultural Trade

Comparing world and U.S. export growth rates by product provides some indication of U.S. competitiveness. Growth rates in total merchandise trade have been greater for the world than for the United States since 1960. As a result, the United States has seen its share of world merchandise trade decline. For all agricultural products, the average real growth in world exports over the 1961-90 period was slightly greater than for U.S. exports. One might hypothesize that U.S. agricultural exports were relatively more competitive (or relatively less noncompetitive) in world markets than for total merchandise exports.

Disaggregating the data by decades and category of agricultural exports provides more information. Over the 1961-90 period, average annual growth rates for U.S. exports were higher than for world exports in both total high-value products and bulk products. Within the high-value product category, the United States has improved its relative position in highly processed products. U.S. highly processed exports fared poorly compared to the world in the 1960s, grew at a slightly faster rate than did world trade in the 1970s, and grew twice as fast as did world trade in those products in the 1980s.

For semiprocessed products, U.S. exports have consistently grown less rapidly than world exports, even though the United States is still a large exporter of these products. There could be many explanations for this, including EC export restitutions on semiprocessed products and Brazilian export tax breaks on soybean oil and meal relative to raw soybeans.

U.S. exports of unprocessed, high-value products grew twice as fast as did world exports over the last three decades. However, the U.S. competitive edge seems to have narrowed in the 1970s and 1980s, compared to the 1960s. U.S. exports of bulk agricultural products grew less rapidly that did world exports of bulk products in the 1960s, much faster in the 1970s, but declined much more rapidly in the 1980s.

Other data indicate that U.S. exports grew at a faster rate than did world trade for processed livestock products (3.6 percent versus 3.2 percent), live livestock (5.3 percent versus 3.0 percent), unprocessed fruits (3.6 percent versus 3.0 percent), bulk grains (2.8 percent versus 2.6 percent), and unprocessed tropical products (2.4 percent versus 1.2 percent).

One interpretation of the data on trade growth rates is that the United States has been relatively more competitive in agricultural products than in all merchandise, relatively more competitive than the rest of the world and gaining in highly processed products, less competitive in semiprocessed products, more competitive but slipping in unprocessed high-value products, and somewhat of a residual supplier for bulk products, especially at levels above the 30 to 35 percent trade shares that characterized the 1960s and 1980s.

Comparative World and U.S. Export Unit Values

The export unit-value for U.S. agricultural exports has consistently been below the overall world average—in 1990 the U.S. average was $318 per metric ton, $523 for the world, and $782 for the EC. Some people have used such statistics to promote the exportation of value-added food products.

The average value per unit of exports reflects the composition of exports and reveals nothing about comparative advantage or competitiveness. Obviously, the United States could greatly increase the average value per ton of exports by dropping bulk product exports, but that would make no sense. Clearly, each category of agricultural products has to compete on its own and the final composition of exports, including average value per ton, is the result. It is more revealing to determine the factors explaining each county's export composition (for example, to what degree is the composition a result of true comparative advantages versus policy- driven market distortions?).

U.S. Exports and Imports

A comparison of the levels and composition of U.S. agricultural exports versus imports lends insight into the types of products in which the United States has the greatest and least comparative advantage or competitiveness. While total agricultural exports exceed imports for the United States, the content of exports contain a higher proportion of bulk products than is the case for imports.

Another way to examine trade balances for competitiveness is to compare imports versus exports to determine the extent to which the United States covers its own needs. The "import cover rate" is simply exports as a percentage of imports. The data in Table 10.3 show the import cover rate for all U.S. merchandise declining over the last three decades. The U.S. imports more processed agricultural products than it exports, but the import cover rate for processed products has strengthened moderately over the last three decades.

While the United States is a net importer of unprocessed high-value prod-

TABLE 10.3 United States Import Cover Rates, 1960s, 1970s, and 1980s
Averages (exports as a percentage of imports)

	1960s	1970s	1980s
Total merchandise	113	86	70
Total agriculture	143	210	202
All HVP	69	89	97
Highly processed	48	42	54
Semi-processed	189	267	269
Unprocessed HVP	17	24	34
Bulk	469	867	1,125

Source: Food and Agriculture Organization

ucts, the import cover rate has been rising, suggesting a growing competitiveness of U.S. products in that market. Also, the U.S. exports far more bulk products than it imports. The interesting point here is that even though the U.S. share of the bulk products market declined in the 1980s, the import cover rate actually rose. The import cover rates for unprocessed, high-value products and bulk products have increased substantially since the 1960s and the import cover rate for highly processed products has improved modestly.

Conclusions and Observations

What can we conclude about the competitiveness of the U.S. food sector from this cursory review of concepts, productivity data, and trade data? Not much that is definitive or "proven." But some general observations can be made.

- Productivity of the U.S. farm production sector is high and continues to grow at rates comparable to that of some other major exporters, but less than others.
- Comparative rates of growth in exports, trends in market shares, and trends in import cover rates suggest that the United States is relatively competitive in bulk commodities, unprocessed high-value products, and some highly processed products, such as processed livestock products.
- The U.S. competitive position has been slipping in semiprocessed food and agricultural products. Overall, the United States appears relatively more competitive in food and agricultural products than in all merchandise.
- The U.S.'s overall comparative advantage in food exports is likely greater than its actual competitive position in world markets because of barriers to trade and policy-related market distortions.

To go beyond these general statements, more work is needed on country agricultural productivity measures for individual countries, estimates of total factor productivity for parts of the food sector beyond the farm, and more detailed analysis of trade data at less aggregated levels than processed, unprocessed, and bulk.

References

Arnade, C. *Productivity of Brazilian Agriculture, Measurement and Uses*. USDA Economic Research Service Staff Report, No. AGES 9219, 1992.

Gullickson, W., and M. Harper. 1987. "Multifactor Productivity in U.S. Manufacturing, 1949-83." *Monthly Labor Review* 110(October 1987): 18-28.

Handy, C., and D. Henderson. "Implications of a Single EC Market for the U.S. Food Manufacturing Sector." In *EC 1992: Implications For World Food and Agricultural Trade*, edited by David Kelch. USDA Economic Research Service Staff Report No. AGES 9133, 1991.

Organization for Economic Cooperation and Development. *Industrial Revival Through Technology*. Paris: OECD, 1988.

Schluter, G., and W. Edmondson. "Exporting Processed Instead of Raw Agricultural Products." USDA Economic Research Service Staff Report No. AGES 89-58, 1989.

11

Assessing the Role of Foreign Direct Investment in the Food Manufacturing Industry

Charles R. Handy and Dennis R. Henderson

Introduction

Modern measures of a country's international competitiveness often incorporate sales from foreign affiliates of home firms (e.g., de la Torre, Horstmann and Markusen, Julius, Kravis and Lipsey, Porter). Foreign affiliation can occur in various ways, e.g., license production to a foreign firm, franchise a foreign firm to market products under the home firm's trademark, acquire a minority interest in a foreign firm, develop a joint venture with a foreign partner, or obtain complete or majority ownership of foreign operations. For purposes herein, foreign direct investment (FDI) refers to investment in a foreign affiliate; the term "foreign affiliate" is used to identify a foreign entity in which a parent firm holds a substantial (but not necessarily majority) interest. Parent firms are referred to as multinational firms (MNCs). In most instances, FDI is measured in terms of sales because asset valuations are difficult to compare over time.

The objectives of this paper are to: (1) set FDI in the context of international activities by food and beverage manufacturers (hereafter referred to as food manufacturers), (2) quantify U.S. in-bound and out-bound FDI in the food manufacturing industries, (3) characterize leading multinational firms (MNCs) with food manufacturing operations in the United States and elsewhere, (4) examine the extent to which FDI is a type of strategic behavior followed by food manufacturing firms, and (5) identify characteristics of food manufacturing firms that are associated with FDI intensity.

International Food Trade in Perspective

International Strategies

There are many alternative strategies that firms can use to enter foreign markets. Some involve considerably more investment in time, money, risk and expertise than do others. In Figure 11.1 these strategies are ordered by the degree of investment and involvement required by home-market firms.

The first three strategies relate primarily to exporting; the remaining three involve varying degrees of direct investment. Most firms initially enter the export market by using foreign agents or brokers. As export sales increase, the next step for many firms is setting up a separate export office or division within their home operations.

Domestic food manufacturers also can pack under contract for a foreign firm. For example, several Japanese manufacturers of soda and fruit drinks contract production of their home brands to American bottlers. This is nearly identical in concept to co-pack operations for private label accounts.

Firms may also choose to have their branded products produced and marketed overseas under a licensing agreement with a foreign firm. While this generally requires no direct investment in foreign production facilities, sizeable outlays may be required to identify appropriate licensees, develop production and marketing procedures, and establish quality control regimes. Of the six strategies, licensing is often the least visible in publicly-available statistical data.

Export Strategies

- Foreign agents and/or brokers
- Domestic export offices
- Co-pack agreements

Foreign Investment Strategies

- License agreements
- Joint ventures
- Foreign subsidiaries

FIGURE 11.1. Strategies to Access Foreign Markets.

Information on licensing is generally not included in trade and investment reports. Licenses are also frequently omitted from company annual reports.

Joint ventures allow a domestic firm to tap into the production, marketing and regulatory expertise of a host-country firm without the expense of acquiring control of a foreign subsidiary. Finally, a manufacturer can acquire or build foreign manufacturing facilities and operate them as a subsidiary. In actual practice, firms often use various combinations of these strategies at the same time.

In this chapter we focus primarily on exports by multinational food manufacturing firms and on their investment in and operation of foreign affiliates.

Trade in Manufactured Foods and Beverages

The value of international trade in products produced by food manufacturing industries world-wide, those industries classified into the 2-digit U.S. Standard Industrial Code 20 (SIC20, Manufactures of Food and Kindred Products), exceeded $205 billion in 1990 (Table 11.1). This is about 3 times the value of world trade in bulk agricultural commodities, represents an estimated 14 percent of the total value of world-wide production of manufactured foods, and reflects a rate of real increase averaging 9.4 percent per year since 1962. By contrast, it is only about one-third as large as the estimated global value of foods produced by foreign affiliates of the world's multinational food manufacturing firms.

Trade in manufactured foods is highly concentrated. Just 19 countries account for 89 percent of all imports (Table 11.2) while 80 percent of all international shipments originate in just 24 countries (Table 11.3). Most of the leading importers are industrialized countries; all are members of the Organization for Economic Co-operation and Development (OECD) except Hong Kong, Korea, Mexico and Singapore. Japan is the leading importer; its share has

TABLE 11.1 World Trade in Manufactured Foods and Beverages, 1962 - 1990 (million U.S.$)

Year	Nominal Value	Value in 1987 Dollars [1]
1962	16,219.9	49,749.7
1967	21,973.3	62,496.5
1972	38,033.8	88,801.3
1977	89,084.7	133,083.6
1982	120,838.7	132,318.4
1987	167,916.1	167,916.1
1990	205,955.6	181,298.9

[1]Based on the Producer Price Index for Finished Consumer Foods.
Source: United Nations.

TABLE 11.2 Leading Importers of Manufactured Foods and Beverages[1]

Country	Share of World Total 1990 (percent)	Percent Change in Share, 1962 - 1990
Japan	12.0	+305.9
Germany	11.8	+0.1
United States	11.7	-50.2
France	8.6	-8.7
United Kingdom	8.6	-58.2
Italy	8.1	+103.2
Netherlands	5.2	+68.8
Belgium/Luxembourg	4.0	+67.4
Spain	3.5	+229.5
Canada	2.6	+0.8
Hong Kong	2.0	+49.1
Denmark	1.6	+33.6
Korea	1.6	+1,765.7
Mexico	1.5	+48.5
Switzerland	1.5	+278.5
Singapore	1.2	-19.3
Greece	1.2	+176.8
Sweden	1.2	-38.0
Portugal	1.0	+226.0
Total of Above	88.9	0

[1]Countries accounting for 1 percent or more of total imports in 1990.
Source: United Nations.

increased rapidly during the past 30 years, recently passing the U.S., whose share has declined by about 50 percent, and Germany, with a stable market share. Korea, Switzerland, Spain, and Portugal have also registered sizeable increases.

There is somewhat greater diversity among the leading supplier countries. About 60 percent of these are OECD members; countries with large agricultural sectors, e.g., Brazil, Thailand, China; newly industrializing countries (NICs), e.g., Taiwan, Korea; and countries with unique agricultural industries, e.g., Colombia, characterize the remainder. The United States is the third largest supplier, following two European countries with highly developed food manufacturing industries, France and the Netherlands. Countries making the largest relative gains are members of the European Community (EC), in particular France, Germany, the Netherlands, and Belgium/Luxembourg. The U.S. share has declined somewhat since the early 1960s; however, the decrease

TABLE 11.3 Leading Suppliers of Manufactured Foods and Beverages to International Markets[1] (Percent)

Country	Share of World Total 1990	Share of World Total 1962
France	9.8	3.9
Netherlands	8.9	5.7
United States	8.5	8.8
Germany (West)	6.7	1.5
United Kingdom	4.3	3.3
Belgium/Luxembourg	4.1	1.3
Denmark	3.9	5.1
Brazil	3.5	6.2
Italy	3.5	2.2
Canada	2.8	3.6
Australia	2.7	4.9
Thailand	2.7	1.3
China	2.6	1.2
Spain	2.1	1.7
Ireland	2.1	1.2
New Zealand	1.7	3.7
Argentina	1.7	4.0
Taiwan	1.6	1.1
Malaysia	1.3	0.7
Indonesia	1.1	1.0
Norway	1.1	1.3
Korea	0.9	0.1
Columbia	0.9	2.5
Mexico	0.9	1.8
Total of Above	79.7	68.0

[1]Countries supplying 0.9 percent or more of total world imports in 1990. *Source:* United Nations.

has been modest when compared to Brazil, Argentina, New Zealand, Australia, and Colombia.

A more detailed examination of U.S. exports of manufactured foods (Table 11.4) reveals a somewhat erratic trend; increasing at a rate above the world average in the early 1960s and again in the early 1980s and at a below-average rate during much of the 1970s and in the mid 1980s. Annual changes vary widely, ranging from increases of 20 percent or more (1979, 1988) to losses of nearly 18 percent (1982). This variability, however, is less pronounced than for bulk agricultural commodities, where annual changes have ranged from an increase exceeding 100 percent (1973) to a decrease of 30 percent (1985). In

TABLE 11.4 United States Exports of Manufactured Foods and Beverages, 1962 - 1991

Year	U.S. Exports ($ million, nominal)	U.S. Exports ($ million, 1987 Dollars)[1]	Annual Change in Real Export Value (percent)	U.S. Share of World Total (percent)
1962	1,206.8	3,701.6	NA	8.8
1963	1,335.0	4,141.2	10.6	8.5
1964	1,658.7	5,130.7	19.3	8.9
1965	1,831.7	5,450.4	5.9	9.6
1966	1,827.1	5,103.8	-6.8	8.7
1967	1,635.8	4,652.6	-9.7	7.8
1968	1,794.6	4,912.8	5.3	8.6
1969	1,939.5	5,008.8	1.9	8.1
1970	2,201.0	5,502.4	9.0	8.0
1971	2,464.1	6,063.5	9.3	8.1
1972	2,787.1	6,507.2	6.8	7.6
1973	3,895.2	7,549.1	13.8	7.7
1974	5,203.1	8,846.9	14.7	8.3
1975	5,130.5	8,048.5	-9.9	7.6
1976	5,688.1	8,948.9	10.1	8.0
1977	6,454.5	9,642.1	7.2	6.7
1978	6,286.3	8,615.2	-11.9	6.8
1979	9,481.4	11,892.5	27.6	7.8
1980	11,942.3	14,152.4	16.0	9.2
1981	12,776.5	14,304.9	1.1	10.2
1982	11,088.4	12,136.2	-17.9	9.4
1983	10,271.2	11,135.6	-9.0	9.0
1984	10,505.2	10,913.9	-2.0	8.6
1985	9,631.3	10,082.4	-8.2	8.0
1986	10,924.0	11,148.0	9.6	7.6
1987	12,707.8	12,707.8	12.3	7.9
1988	15,956.6	15,522.0	22.2	8.6
1989	16,960.5	15,646.2	0.8	8.9
1990	17,490.4	15,396.5	-1.6	8.5
1991	20,084.4	17,726.7	15.1	NA

[1] Based on the Producer Price Index for Finished Consumer Foods.

Source: United Nations through 1990; U.S. Bureau of Census thereafter.

absolute terms, exports of manufactured foods account for about half of the value of all U.S. food and agricultural exports; in contrast, on a world-wide basis, manufactured foods account for about 75 percent of the total.

The composition of U.S. manufactured food exports is illustrated by Figure

11.2; meats account for the single largest component. When viewed in the context of a trade balance (Table 11.5), a narrowing gap is evident; current expectations are, the United States will record a surplus of manufactured food exports over imports in 1992.

U.S. Exports by Multinational Firms

While U.S. exports of manufactured foods have grown rapidly since 1985, export propensity is relatively low. That is, exports account for a relatively small share of the total value of U.S. output. In 1991, exports totaled 5.2 percent of all shipments from U.S. plants, up slightly from 4.7 percent in both 1988 and 1989. To gain perspective on the contribution by multinational firms to U.S. exports, data from two sources are analyzed. Aggregate data at the SIC20 level come from the U.S. Department of Commerce, Bureau of Economic Analysis (BEA 1991a). Additionally, data on 71 large U.S. food manufacturers are used for firm-level analysis; these data have been compiled in an unpublished data base by the Economic Research Service (ERS). The latter includes information

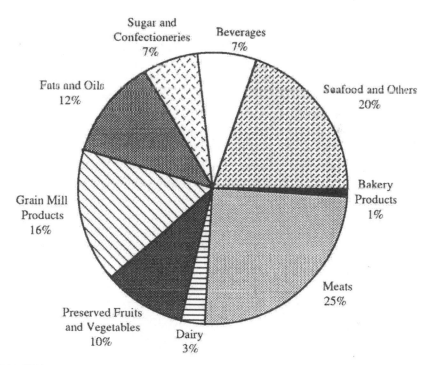

FIGURE 11.2 U.S. Processed Food Exports, 1991

TABLE 11.5 U.S. Exports and Imports of Manufactured Foods, 1988 - 1991 ($ million)

Year	Exports	Imports
1988	16,414.2	19,399.9
1989	17,111.7	19,681.8
1990	18,585.5	20,876.7
1991	20,084.4	20,806.7

Source: U.S. Bureau of Census.

on total firm sales, sales of processed foods, sales of processed foods from U.S. operations, exports from U.S. operations, processed food sales from foreign affiliates, and the number and location of foreign food manufacturing plants.

United States MNCs classified as food manufacturers export between 3 and 4 percent of their sales (BEA 1991a). Therefore, food manufacturing MNCs exhibit a somewhat lower export propensity than do all food manufacturing firms. In 1989, 64 U.S.-based MNCs in the ERS data base that are classified as food manufacturers reported export sales of $7.9 billion, 4.2 percent of their total sales. As this includes exports of non-food products, this is an upper bound on exports of food. After adjusting for non-food products, food exports of these firms are estimated to be about $5.6 billion, equivalent to 3 percent of their U.S. turnover. At this level, U.S.-based MNCs account for about one-third of total U.S. exports of manufactured foods.

In addition to the exports shipped from home-country plants of U.S.-based MNCs, U.S. operations of non-U.S. MNCs also originate exports. U.S. affiliates of foreign firms exported $2.0 billion in processed foods in 1989 (BEA 1991a). Thus, foreign-owned food manufacturing facilities in the United States accounted for 12 percent of all U.S. exports of manufactured foods and beverages. Combining the export shares from U.S. operations of both U.S. and non-U.S. MNCs brings the share of U.S. manufactured food exports originated by multinational firms to about 45 percent.

A more detailed export profile of U.S.-based MNCs can be obtained from the ERS data base. This includes 34 MNCs operating in both 1988 and 1990 (Table 11.6). On average, these large firms have export propensities significantly lower than the mean value for all food manufacturers. In 1988, exports of this group of MNCs totaled $2.9 billion, accounting for 2.6 percent of their total sales from U.S. operations. By comparison, exports for all U.S. food manufacturers averaged 4.7 percent of total sales. Recently, however, export activity has increased dramatically for many MNCs. Between 1988 and 1990, exports by this group rose 59.5 percent, to $4.6 billion, raising the export propensity for the group of MNCs as a whole to 3.5 percent. The share of total U.S. exports generated by these 34 firms rose from 18 percent in 1988 to 25 percent in 1990. All but seven increased export shipments during this period.

TABLE 11.6 Food and Beverage Exports of U.S. Firms with Foreign Affiliates, 1988 and 1990

	1988		1990		
	Exports ($ 000)	Share of U.S. Food Sales (percent)	Exports ($ 000)	Share of U.S. Food Sales (percent)	Percent Change, 1988 - 1990
Philip Morris/Kraft-General Foods	263,541	1.5	1,232,265	5.5	367.6
Archer Daniels Midland	978,968	16.5	954,965	14.7	-2.5
Con Agra	215,456	3.0	536,239	3.6	148.9
Anheuser Busch	282,378	3.5	445,878	4.6	57.9
Tyson	152,562	4.4	184,700	5.0	21.1
Coca Cola	93,932	2.6	118,000	3.0	25.6
Proctor & Gamble	124,446	4.3	115,799	3.6	-7.0
General Mills	73,940	2.2	95,338	2.3	28.9
Heinz	61,051	2.0	90,718	2.4	48.6
Chiquita Brands	85,835	3.6	80,046	3.8	-6.7
Ralston Purina	39,865	1.1	74,670	2.0	87.3
Hershey's	39,372	2.0	68,000	2.7	72.7
Sara Lee	38,369	0.7	64,527	1.3	68.2
McCormick	61,552	6.4	59,500	6.0	-3.3
Kellogg's	42,679	1.7	54,789	1.8	28.4
Brown Forman	51,384	6.5	46,224	7.0	-10.0
MM/Mars	44,550	1.0	45,000	1.0	1.0
Universal Foods	35,975	5.5	44,476	5.8	23.1
CPC International	30,652	1.5	39,477	1.7	28.8
American Brands	13,818	3.5	39,207	4.2	183.7
Castle & Cooke	1,360	0.1	35,506	2.0	2,510.7
Borden	21,872	0.6	35,100	0.8	60.5
International Multifoods	12,056	1.0	26,168	1.5	117.1
Quaker Oats	14,619	0.5	23,644	0.7	61.7
PepsiCo	21,428	0.4	23,304	0.3	8.8
Campbell Soup	30,000	0.7	22,401	0.5	-25.3
Pet Inc.	24,335	1.3	19,328	1.3	-20.6
RJR/Nabisco	14,000	0.2	17,000	0.3	21.4
American Home Products	12,300	2.0	16,247	2.0	32.1
Gerber Foods	5,385	1.1	10,552	1.9	96.0
Curtice-Burns	11,189	1.8	9,834	1.1	-12.1
Smucker's	7,350	2.1	9,746	2.5	32.6
Wm. Wrigley	4,851	0.9	6,208	0.9	28.0
Clorox	1,266	0.6	1,544	0.6	22.0
Total	2,912,336	2.6	4,646,200	3.5	59.4

Source: Original data.

Export activity varies widely among these U.S. MNCs. Kraft-General Foods, owned by Philip Morris, is the largest exporter, with processed food exports of $1.2 billion in 1990, up 368 percent from 1988. Archer Daniels Midland is the second largest, with exports of $955 million, followed by ConAgra ($536 million) and Anheuser Busch ($446 million). In all, seven of the 34 firms had export sales exceeding $100 million in 1990.

At the other end of the spectrum, seven of the U.S. MNCs had export propensities of less than one percent. Borden's exports in 1990 totaled just $35 million, only 0.8 percent of sales. Reported exports were 0.7 percent of sales for Quaker Oats, 0.5 percent of sales for Campbell Soup, and 0.3 percent of sales for both PepsiCo and RJR/Nabisco. Even though each of these firms has large and sophisticated international management and marketing staffs, that international expertise is used primarily for commercial activities other than exporting products produced in the United States.

Foreign Direct Investment in the U.S. Food Manufacturing Sector

Investment Abroad by U.S. Food Manufacturers

Most large food manufacturers rely more heavily on foreign direct investment than on exports as their major strategy to access foreign markets. In 1989, U.S. MNCs held at least 10 percent equity in 720 food manufacturing affiliates (BEA 1991b). Sales from those foreign affiliates grew from $39 billion in 1982 to $69 billion in 1989, an average annual change of 11 percent (Table 11.7). While more recent sales data from foreign affiliates are not yet available, foreign capital expenditures by U.S. MNCs grew 10 percent in 1990 and another 18 percent in 1991. This is one indication that affiliate sales continue to expand at about the 10 to 11 percent annual rate.

Most U.S. FDI is placed in developed countries. Sales by affiliates in Europe, Canada, and Japan account for 75 percent of all U.S. affiliate sales. However, affiliate sales in Europe have a more rapid average annual change (16.6 percent) than in either Canada (7.9 percent) or Japan (5.4 percent).

TABLE 11.7 Value of Shipments from Foreign Food Manufacturing Affiliates of U.S. Firms, 1982 -1989

	1982 ($ million)	1987 ($ million)	1989 ($ million)	Average Annual Change, 1982 - 1989 (percent)
Total, all countries	39,023	50,067	69,033	11.0
Europe	18,974	29,044	40,985	16.6
Canada	5,258	5,522	8,148	7.9
Japan	2,363	4,442	3,250	5.4

Source: Bureau of Economic Analysis, August 1991.

Firm-level data compiled by ERS provide the basis for a more detailed analysis of FDI behavior by the 34 large U.S. food manufacturing MNCs. For these firms as a group, sales from foreign affiliates increased from $39.6 billion in 1988 to $41.1 billion in 1990, a 3.7 percent gain (Table 11.8). Seven of firms in the group reported declines in sales from foreign affiliates; two had major divestments of foreign operations. RJR/Nabisco sold a large share of its European food manufacturing operations and Chiquita Brands sold its food manufacturing operations in the UK. Largely as a result of these and other divestitures, the foreign affiliate share of total food sales for all 34 firms declined from 26.8 percent in 1988 to 24.2 percent in 1990. Preliminary data, however, indicate that the foreign affiliate share increased sharply, to 29 percent, in 1991.

Of the U.S. MNCs, 13 report more than $1 billion in sales by their foreign affiliates in 1990. The largest, Kraft-General Foods, generated 24.9 percent of its total turnover of manufactured foods, $7.4 billion, from its foreign affiliates. This firm is also the largest U.S. exporter of manufactured foods. Two firms, Coca-Cola and CPC International, report that about 60 percent of their world-wide food sales originate with their foreign operations. An additional seven firms originate 30 percent or more of their sales from foreign affiliates.

In 1990, the U.S. parent firms in the ERS data base had 734 food manufacturing plants abroad. While most FDI is in developed countries, 215 (29 percent) of these plants were located in developing countries (excluding the former USSR, Eastern Europe, and China); 23 of the firms had affiliates in developing countries. CPC International is the largest investor in developing countries, with 44 plants, followed by Ralston Purina with 34 plants. Sales from U.S. affiliates in Mexico grew from $1.6 billion in 1987 to an estimated $4.1 billion in 1991.

Eastern and Central Europe are relatively new and potentially fast-growing markets for U.S. FDI in food manufacturing. From company annual reports and reports in trade journals, we estimate that U.S. parents had 22 affiliates in this area as of mid-1992. Examples include Sara Lee's acquisition of Compack Trading, the third largest food manufacturing firm in Hungary. Gerber, likewise, acquired Alima SA, the largest infant food and juice firm in Central Europe, from Poland's Ministry of Privatization. In addition, U.S. parents have nine food processing affiliates in the former USSR and nine plants in China.

Foreign affiliates of U.S. MNCs, as a group, are more export oriented than are their parent firms (BEA 1991b). While U.S. parents export about 3.5 percent of their domestic production, exports account for an average of 19 percent of the sales of their foreign affiliates. But, affiliate export propensities vary widely, closely matching the export behavior of their host country firms. For example, U.S. affiliates in Canada are very home-country oriented, exporting only about 5 percent of their production. By contrast, U.S. affiliates in Europe export nearly 25 percent of their turnover, with 85 percent of those exports shipped to other European countries.

TABLE 11.8 Sales by Foreign Affiliates of U.S. Food Manufacturing Firms, 1988 and 1990

	1988		1990		
	Sales ($000)	Share of Total Food Sales (percent)	Sales ($000)	Share of Total Food Sales (percent)	Percent Change, 1988 - 1990
Philip Morris/Kraft-General Food	8,556,063	33.2	7,388,000	24.9	-13.7
Coca Cola	4,319,234	54.0	6,260,600	61.4	45.0
CPC International	2,656,500	56.5	3,458,900	59.8	30.2
PepsiCo	2,030,000	24.9	3,071,000	26.5	51.3
MM/Mars	2,295,000	34.0	2,362,500	35.0	2.9
Heinz	2,191,647	41.8	2,305,802	37.9	5.2
Kellogg's	1,762,216	40.5	2,137,600	41.3	21.3
Sara Lee	1,739,842	23.9	1,935,401	28.5	11.2
Campbell Soup	1,503,304	26.5	1,747,850	28.2	16.3
Quaker Oats	1,584,200	35.1	1,652,900	32.9	4.3
Ralston Purina	1,140,000	24.1	1,600,050	30.0	40.4
Borden	1,480,200	27.5	1,220,488	21.7	-17.6
Castle & Cook	917,069	40.4	1,016,690	36.4	10.9
RJR/Nabisco	2,981,000	30.2	669,990	11.5	-77.5
Anheuser-Busch	514,974	6.0	608,016	6.0	18.1
Con Agra	330,545	4.4	520,918	3.4	57.6
International Multifoods	492,376	29.0	447,351	20.4	-9.1
Wm. Wrigley	320,782	36.0	433,701	38.6	35.2
Pet Inc.	517,418	22.0	409,904	22.0	-20.8
General Mills	311,466	8.3	375,185	8.3	20.5
Archer Daniels Midland	183,555	3.0	341,059	5.0	85.8
McCormick	171,400	15.2	256,500	20.4	49.7
Hershey's	199,401	9.2	207,067	7.6	3.8
Tyson	0	0	145,000	3.8	
Chiquita Brands	1,144,483	32.7	111,174	5.0	-90.3
Proctor & Gamble	73,705	2.5	82,618	2.4	12.1
Universal Foods	64,755	9.0	75,500	9.0	16.6
American Brands	2,000	0.5	70,371	7.0	3,418.6
Gerber Foods	36,931	7.0	41,810	7.0	13.2
American Home Products	31,224	4.8	40,958	4.8	31.2
Brown Forman	46,445	5.5	38,506	5.5	-17.1
Smucker's	16,843	4.6	31,964	7.6	89.8
Curtice-Burns	16,240	2.5	22,234	2.5	36.9
Clorox	4,612	2.0	5,432	2.0	17.8
Total	39,635,430	26.8	41,093,039	24.2	3.7

Source: Original data.

Direct trade links with affiliates are rather small. Manufactured food exports from the United States to foreign affiliates of U.S. firms totaled $2.2 billion in 1989. This accounts for only 13 percent of total U.S. exports. Yet, exports to affiliates are more than twice as large as are U.S. imports from affiliates. In 1989, only 2 percent of affiliate sales, $900 million, was shipped to U.S. parents.

While the volume of trade between the United States and foreign affiliates of U.S. MNCs is modest, most of that trade is intra-firm. Of the $2.2 billion in U.S. exports to affiliates, $1.7 billion (71 percent) was shipped by U.S. parents. Likewise, of the $900 million in U.S. imports shipped from foreign affiliates of U.S. firms, 85 percent is shipped to U.S. parent firms.

Recognizing that much research remains to be done and that generalizations can be risky, the evidence at this point suggests that, while large multinational firms appear to view exports and FDI as alternative international marketing strategies, U.S. investment abroad is not necessarily trade diverting. Not only have total U.S. exports of manufactured food increased appreciably in recent years, the share of those exports originated by U.S. MNCs rose by 7 percentage points between 1988 and 1990. Reed and Marchant (1992) and Lipsey (1991) have reported similar findings.

Investment in the United States by Foreign Food Manufacturers

Sales from foreign-owned food manufacturing operations in the United States are considerably smaller than are those from U.S.-owned affiliates abroad, but have been expanding at a more rapid pace. Sales by these foreign-owned affiliates in the United States reached $41.1 billion in 1989, up from $14.8 billion in 1982, an average annual change of 25.3 percent (Table 11. 9). Thus, during the 1980s, sales from U.S. affiliates of foreign firms grew more rapidly than did sales from foreign affiliates of U.S. firms. However, foreign investment in the United States has slowed considerably since 1989. Sales from foreign-owned affiliates increased about 9 percent in 1990. Annual outlays to acquire or establish U.S. food manufacturing enterprises have also decreased recently. After peaking at $6.5 billion in 1989, annual in-bound FDI outlays dropped to $1 billion in 1990 and fell to $757 million in 1991. About 90 percent of all outlays

TABLE 11.9 Value of Shipments from U.S. Food Manufacturing Affiliates of Foreign Firms, 1982 -1989

	1982 ($ million)	1987 ($ million)	1989 ($ million)	Average Annual Change, 1982 - 1989 (percent)
Total, all countries	14,847	22,862	41,120	25.3
Europe	10,527	17,967	30,785	27.5
Canada	2,218	3,174	5,562	21.5
Japan	564	612	1,644	27.4

Source: BEA, October 1991.

was for acquiring existing U.S. food manufacturers rather than for establishing new operations.

Almost all foreign investment in U.S. food manufacturing comes from developed countries. European firms alone account for 75 percent ($30.8 billion in annual sales). U.S. affiliates of Canadian firms accounted for another 14 percent ($5.6 billion). Sales from U.S. affiliates of Japanese firms increased at an average annual rate of 27 percent from 1982 to 1989, but at $1.6 billion, account for just 4 percent of all foreign affiliate sales.

Regarding the trade-orientation of foreign-owned food manufacturing affiliates in the United States, in 1989 the $2 billion in exports from these operations accounted for 5 percent of their total turnover. Thus, their export propensities are only marginally higher than for the U.S. operations of U.S. MNCs. Exports from foreign-owned firms account for about 12 percent of total manufactured food exports from the United States. Imports are somewhat larger, amounting to $2.7 billion, 6.5 percent of total affiliate sales. Foreign-owned firms tend to be more import-oriented than are U.S. food manufacturers as a whole. The net result is, U.S. affiliates of foreign firms tend to export more and import more relative to their total sales than do U.S.-based firms. The differences, however, are not large.

Of the $2 billion exported by foreign affiliates, nearly half (48 percent) was shipped to their parent firms. By region, 72 percent of these exports went to Europe; 16 percent to Japan. Intra-firm trade also accounts for a high share of imports. Foreign parents accounted for more than half (58 percent) of all manufactured food imports received by their U.S. affiliates. Of all imports shipped to foreign affiliates in the United States, 58 percent originated in Europe; 29 percent in Canada.

In summary, U.S. MNCs rely far more heavily on foreign investment strategies than on exports to access foreign markets. Yet, the export propensities of MNCs have increased in recent years. Foreign-owned affiliates in the United States trade a higher share of their total turnover in international markets than do the domestic operations of U.S. MNCs. As with U.S. investment abroad, inward FDI in U.S. food manufacturing operations appears to be mildly trade augmenting rather than trade diverting. But, the net effect is small.

Characteristics of Multinational Food Manufacturing Firms

World Panel of Food Manufacturers

To gain insights into the organization and performance of multinational food firms and to examine MNCs for characteristics that distinguish them from home-oriented firms, data were drawn from a sample of more than 600 of the world's leading firms with food manufacturing operations. These firms were identified from the CIFAR listing of food and beverage firms, and data were drawn from corporate annual reports and a variety of commercially-available sources (Hirschberg et al.).

From this sample, a panel was constructed consisting of 144 food manufacturing firms for which geographical data on investment and operations are available. This panel is not necessarily a representative sample of food manufacturers world-wide; however, it does include many of the leading firms that are headquartered in the established industrialized countries (Table 11.10). Of these firms, 93 are American or British; unfortunately, firms in a number of countries have little history of reporting investment or operating information on a geographic basis. Nonetheless, firms in this panel provide, in toto, an insightful view of multinational behavior in the food manufacturing sector. While any inter-relationships between organization and performance found among firms in this panel may not "prove the case", they can be interpreted as strongly suggestive of viable hypotheses.

To provide a measure of the extent to which this panel represents the food

TABLE 11.10 Sample of Leading World Food Manufacturing Firms: Nationality

Home Country	Number of Firms	Home Country	Number of Firms
Australia	6	New Zealand	1
Canada	6	Netherlands	6
Denmark	3	Norway	1
Finland	2	Sweden	1
France	10	Switzerland	3
Germany	3	South Africa	2
Ireland	2	United Kingdom	23
Japan	5	United States	70

Source: Original data.

TABLE 11.11 Shipments by Leading World Food Manufacturing Firms: Geographic Coverage, 1989/1990

	Number of Firms	Annual Shipments (million U.S.$)	Share of Region Total Shipments (percent)
North America	76	186,034	49.8
European Community	47	116,525	35.4
Japan	5	11,282	16.7
Other OECD	14	53,637	82.5
Rest of World	2	2,217	NA

Source: Original data; OECD; authors' estimates.

manufacturing industry world-wide, its share of total output of manufactured food is estimated (Table 11.11). It appears that, excepting Japan, reasonable coverage of the industry is reflected throughout the OECD countries; practically none elsewhere.

Descriptive statistics on this panel of firms are provided in Table 11.12. As a group, these are large firms; in the 1989/1990 base period annual shipments or turnover averaged in excess of $3.7 billion, total assets averaged more than $3.4 billion per firm, and firms employed an average of 22,500 workers. Of their annual turnover, 67 percent is manufactured food and beverage products; on average, 63 percent of these products are manufactured in the firms' home country; the remaining 37 percent is produced abroad. For the group as a whole, food operations account for 74 percent of the firms' total operating income, substantially exceeding the 67 percent product share. Thus, even though many of the firms are diversified, food manufacturing appears to be their predominant business.

Of the 144 firms, 118 report foreign operations; the remaining 26 report operations only in their home market. Firm size appears to be associated with the extent to which a firm is multinational; for example, the 118 MNCs averaged $2.8 billion in annual food turnover compared to $2.5 billion for the entire panel; dividing the panel into halves based on share of total shipments originating in foreign plants shows annual turnover averaging $3.5 billion for those in the "upper half" (foreign shipments exceeding 17 percent of total) compared to $1.5 billion for those in the "lower half". Similar size comparisons exist when measured in terms of assets and employees.

Comparing the U.S. and non-U.S. firms in the panel reveals a number of similarities, along with some distinctions. In terms of average annual turnover of food products and number of employees, the firms are nearly equal. On the basis of asset value, U.S. firms exceed the average size of other firms by 64 percent; on the basis of specialization in food products and share of food shipments originating abroad, non-U.S. firms exceed their U.S. counterparts 78 percent to 58 percent, and 52 percent to 20 percent, respectively.

A comparison of non-U.S. MNCs that have operations in the United States with those that do not reveals modest differences (Table 11.13). Overall, those firms with no U.S. affiliates hold a somewhat larger percentage of total investment in foreign operations than do those with U.S. affiliates. However, there is no appreciable difference in the share of total shipments generated abroad. Further, the ratio of shipments to assets appears to be greater for U.S. affiliates of non-U.S. MNCs than for their other (non-U.S.) foreign affiliates. These observations suggest that those firms with U.S. affiliates concentrate a greater proportional effort in one foreign market, i.e., the U.S., than do other non-U.S. MNCs.

TABLE 11.12 Characteristics of Leading World Food Manufacturing Firms, 1989/1990 (Million U.S.$ except where noted)

	All 144 Firms in Sample	118 Firms with Foreign Shipments > 1% of Total	72 Firms with Highest % Foreign Shipments	72 Firms with Lowest % Foreign Shipments	74 Non-U.S. Firms	70 U.S. Firms
Food and Beverage Operations						
Value of Shipments						
Mean	2,502.1	2,812.7	3,528.4	1,475.8	2,509.4	2,494.4
High	28,103.7	28,103.7	28,103.7	10,185.3	28,103.7	26,368.0
Low	100.7	100.7	113.0	100.7	100.7	134.2
Operating Income						
Mean	325.6	342.9	401.9	189.3	245.7	468.3
High	1,995.6	1,995.6	1,995.6	1,301.6	1,121.9	1,995.6
Low	-80.9	-80.9	-80.9	-4.8	0.9	-80.9
Value of Shipments from Home Country Operations						
Mean	1,583.6	1,692.2	1,798.0	1,369.2	1,197.9	1,991.3
High	21,110.8	21,110.8	21,110.8	9,258.4	4,783.5	21,110.8
Low	42.7	42.7	42.7	96.8	42.7	129.3
Value of Shipments from Foreign Operations						
Mean	918.5	1,120.5	1,730.4	106.6	1,311.5	503.2
High	27,568.4	27,268.4	27,568.4	1,046.6	27,568.4	5,257.2
Low	0	3.3	19.9	0	0	0
Foreign Shipments as a Percent of Total Shipments						
Mean	36.71	39.84	49.04	7.23	52.26	20.17
High	98.10	98.10	98.10	16.39	98.10	71.11
Low	0	1.10	17.26	0	0	0
Consolidated Operations						
Value of Shipments						
Mean	3,722.1	4,169.4	4,622.0	2,822.2	3,197.6	4,276.6
High	39,011.0	39,011.0	39,011.0	24,081.0	29,273.1	39,011.0
Low	140.0	140.0	223.9	140.0	159.6	140.0
Operating Income						
Mean	440.7	490.8	577.5	239.3	312.9	700.5
High	6,789.0	6,789.0	6,789.0	2,530.0	3,106.1	6,789.0
Low	-4.8	1.0	18.3	-4.8	1.0	-4.8
Net Income						
Mean	199.2	221.6	276.2	78.3	165.3	269.9
High	2,946.0	2,946.0	2,946.0	767.2	1,799.2	2,946.0
Low	-1,149.0	-1,149.0	-980.2	-1,149.0	-980.2	-1,149.0
Total Assets						
Mean	3,442.1	3,860.9	4,041.7	2,462.2	2,833.8	4,658.7
High	38,528.0	38,528.0	38,528.0	36,412.0	21,576.4	38,528.0
Low	35.7	35.7	159.4	35.7	35.7	174.6
Number of Employees						
Mean	22,562	25,699	29,959	15,271	20,963	24,139
High	266,000	266,000	266,000	90,138	196,940	266,000
Low	335	335	1,000	335	335	600

Source: Original data.

TABLE 11.13 Operating Characteristics of Non-U.S. Multinational Food Manufacturing Firms in the United States and Elsewhere, 1989/1990[1] (percent)

	Firms with U.S. Operations	Firms with no U.S. Operations
Assets		
U.S. Assets/Total Assets	22.6	NA
U.S. Assets/Foreign Assets	55.5	NA
Foreign Assets/Total Assets	40.6	44.7
Shipments		
U.S. Shipments/Total Shipments	24.7	NA
U.S. Shipments/Foreign Shipments	65.2	NA
Foreign Shipments/Total Shipments	37.9	36.9
Earnings		
U.S. Net Income/U.S. Assets	7.1	NA
Foreign Net Income/Foreign Assets	10.3	9.3
Total Net Income/Total Assets	9.0	9.4

[1] Sub-sample of 39 firms.
Source: Original data.

Intensity of Foreign Direct Investment

To get a sense of the extent of multinationality among the world's leading food manufacturers, the panel was divided into U.S. and non-U.S. firms; each group was then separated into multinational firms, i.e., those with foreign operations, and others, i.e., those with no foreign operations (Table 11.14). The multinational firms in each group were then sub-divided into "foreign-oriented" and "home-oriented" classifications, the dividing line being the average

TABLE 11.14 Average Size of Leading Food Manufacturing Firms 1989/1990 (Values in Million U.S.$)

	Number of Firms in Sample	Value of Food Shipments	Consolidated Value of Shipments	Total Assets	Number of Employees
U.S. Firms					
With Foreign Operations	50	3,007.0	5,211.4	6,026.7	30,410
Without Foreign Operations	20	1,212.9	1,939.7	1,549.7	8,459
Non-U.S. Firms					
With Foreign Operations	68	2,669.7	3,403.2	3,040.6	22,018
Without Foreign Operations	6	691.7	866.7	559.2	7,465

Source: Original data.

share of shipments from foreign operations as a percent of total shipments for each group (Table 11.15). Based upon these distinctions, 50 of the 70 U.S. firms are MNCs; 16 of the U.S. MNCs are classified as foreign-oriented and 34 as home-oriented. For the non-U.S. firms, 68 are MNCs; 22 of these are considered foreign-oriented and 46 home-oriented.

The relationship between size and multinationality, demonstrated by others for U.S. food manufacturers (Connor and Murphy, Horst), appears to be confirmed regardless of firm nationality (Table 11.14). While the number of non-U.S. firms in the panel with no foreign operations is small and, quite likely, not representative, the association between size of firm, regardless of measure, and extent of foreign operations is pronounced for both groups. Even so, these observations offer no insight into the direction of causation, i.e., are MNCs large because they are multinational, or are large firms multinational because they are large?

Some insights into the direction of the size-multinationality relationship can be gleaned from indicators of FDI intensity (Table 11.15). The first "intensity" measure is, share of total assets invested abroad. As expected, this share is significantly higher for foreign-oriented MNCs: 37.3 percent for foreign-oriented U.S. firms and 61.8 percent for foreign-oriented non- U.S. firms compared to 20.0 percent and 35.5 percent, respectively, for the home-oriented MNCs.

TABLE 11.15 Intensity of Foreign Investment by Leading Multinational Food Manufacturing Firms, 1989/1990

	Foreign Assets as a Percent of Total[1]	Average Number of Foreign Affiliates[2]
All Firms	36.5	51.9
U.S. Firms	33.6	38.9
Foreign-Oriented[3]	37.3	34.6
Home-Oriented[4]	20.0	45.3
Non-U.S. Firms	40.9	65.8
Foreign-Oriented[5]	61.8	57.1
Home-Oriented[6]	35.5	71.6

[1] Sub-sample of 53 firms.

[2] Sub-sample of 29 firms.

[3] Firms with shipments from foreign operations greater than the average for all U.S. multinational firms: 23.4 percent of total shipments.

[4] Firms with shipments from foreign operations less than 23.4 percent of total shipments.

[5] Firms with shipments from foreign operations greater than the average for all non-U.S. multinational firms: 53.5 percent of total shipments.

[6] Firms with shipments from foreign operations less than 53.5 percent of total shipments.

Source: Original data.

The second "intensity" measure is a count of foreign affiliates. For the sub-sample of firms that identify their foreign affiliates, the average number of such affiliates is 52; strong indication that these are large, complex organizations. On average, non-U.S. MNCs report a greater number of foreign affiliates (66) than do U.S. MNCs (39), numbers that are consistent with both the greater relative importance of foreign assets and the larger share of foreign- originated shipments for the non-U.S. firms.

Of particular interest, however, are the observations that foreign-oriented firms (both U.S. and non-U.S.) have significantly fewer foreign affiliates than do the home-oriented MNCs (35 vs. 45 for U.S. firms; 57 vs. 72 for non-U.S. firms). This implies that foreign-oriented MNCs concentrate their foreign ventures into a relatively small number of (implicitly) larger operations than do their home-oriented counterparts. By conjecture, this suggests that those MNCs with the most extensive foreign operations focus their attention on a limited number of endeavors where they may have particular competitive strengths or advantages. By contrast, those MNCs with a less extensive foreign orientation appear to have more diverse, possibly less focused operations. We will subsequently return to this theme, arguing that these findings are consistent with the theorem that firms become multinational in order to exploit intangible assets.

Specialization and Internationalization

It has been suggested that firm diversity is positively related to propensity to invest abroad (Connor). The essence of the argument is, diversified firms, because of unique product portfolios, can take advantage of unique combinations of industry-specific inducements to FDI, e.g., stable or growing demand, existence of open distribution systems, availability of market information. To examine this possible relationship, specialization and international ratios were calculated for panel firms (Table 11.16). Specialization in this case is the inverse of diversification, measured as the ratio of the value of food shipments to turnover of all products.

For U.S. firms, those with foreign affiliates appear to be somewhat less specialized (or more diversified) than are those with home operations only; the evidence from non-U.S. firms is not convincing. Further, internationalization of food operations by diversified firms does not appear to be appreciably different from their international profile across all products.

The international ratio incorporates exports as well as shipments from foreign affiliates. Even though non-U.S. firms are substantially more oriented toward foreign sales than are U.S. firms, the international ratios for firms with no foreign operations average about 20 percent of those for MNCs regardless of firm nationality. By implication, there is little to distinguish between U.S. and

TABLE 11.16 Specialization and Internationalization of Leading Food Manufacturing Firms, 1989/1990

	Specialization Ratio[1]	Foreign Shipments as a Percent of Total		International Ratio[2]
		Food	All	
U.S. Firms				
With Foreign Operations	57.7	23.4	25.2	26.1
Without Foreign Operations	62.5	0	2.2	5.2
Non-U.S. Firms				
With Foreign Operations	78.5	53.5	43.5	54.3
Without Foreign Operations	79.8	0	0.1	12.2

[1] Food shipments as a percent of total shipments.
[2] Exports from home country plus shipments from foreign operations as a percent of total shipments.
Source: Original data.

non-U.S. firms in terms of exports as a share of their total international market exposure.

Yet, export activity varies significantly between U.S. and non-U.S. firms and between MNCs and firms with no foreign affiliates (Table 11.17). The export propensity of non-U.S. firms is 3.5 times larger than for U.S. firms; another reflection of the greater overall importance of international markets to the former. Regardless of nationality, MNCs demonstrate a lower propensity to export than do other firms, implying that FDI and exports are substitute

TABLE 11.17 Exports of Leading Food Manufacturing Firms, 1989/1990

	Export Propensity[1]
U.S. Firms	3.1
With Foreign Operations	2.7
Without Foreign Operations	5.2
Non-U.S. Firms	11.1
With Foreign Operations	10.8
Without Foreign Operations	12.3

[1] Exports as a percent of total shipments.
Source: Original data.

activities. However, the substitute nature of these activities appears to be substantially greater for U.S. MNCs than their non-U.S. counterparts.

Strategic Value of Foreign Direct Investment

Given that considerable variability is observed in FDI activity among the world's leading food manufacturing firms, it is reasonable to assume that such activity is not necessarily essential to the economic survival of such firms as a class; rather it is a strategic response to perceived profit opportunities. For the panel of firms and sub-groups thereof, available data allow specification of profits in terms of net income as a percent of total assets, i.e., a proxy for rate of return on borrowed and invested capital (ROI).

Comparing all MNCs with firms that have no foreign affiliates reveals that, as a group, the latter are associated with a measurably higher ROI than the former, implying that FDI comes at a cost in the profit account (Table 11.18). This relationship appears to hold for both U.S. and non-U.S. firms; while the reported difference is greater for non-U.S. firms, it may be biased due to the small number of such firms in the sample that have no foreign operations. Further, it appears that ROI for U.S. firms is somewhat below that for non-U.S. firms; however, this may reflect different accountancy practices in the United States and other countries.

The appearance of lower ROI for MNCs than for their non-multinational counterparts raises the issue of motivation for FDI in food manufacturing. However, sub-dividing the MNCs into the classifications "foreign-oriented" and "home-oriented", as defined earlier, is particularly revealing (Table 11.19). In this case, the groups of foreign-oriented MNCs show appreciably higher ROI than do the home-oriented multinationals, the differences being particularly pronounced for the U.S. firms. This suggests that, those firms with a substantial

TABLE 11.18 Earnings of Leading World Food Manufacturing Firms, 1989/1990

	Net Income as a Percent of Assets, Total for Group
All 144 Firms	5.79
118 Multinational Firms	5.74
26 Firms w/o Foreign Operations	6.61
All 74 Non-U.S. Firms	5.91
68 Multinational Firms	5.88
6 Firms w/o Foreign Operations	7.92
All 70 U.S. Firms	5.63
50 Multinational Firms	5.55
20 Firms w/o Foreign Operations	6.36

Source: Original data.

TABLE 11.19 Earnings of Leading Multinational Food Manufacturing Firms, 1989/1990

	Net Income as a Percent of Assets, Total for Group
All Multinational Firms	
35 Foreign-Oriented Firms [1]	8.08
83 Home-Oriented Firms [2]	4.42
Non-U.S. Multinational Firms [3]	
22 Foreign-Oriented Firms	6.03
46 Home-Oriented Firms	5.79
U.S. Multinational Firms	
16 Foreign-Oriented Firms	10.67
34 Home-Oriented Firms	3.69

[1] Firms with shipments from foreign operations greater than the average for all multinational firms in the sample: 39.8 percent of total shipments.
[2] Firms with shipments from foreign operations less than 39.8 percent of total shipments.
[3] For definitions of foreign-oriented and home-oriented non-U.S. and U.S. multinational firms, see Table 11.15.
Source: Original data.

commitment to foreign operations earn correspondingly higher returns than do other MNCs; moreover, MNCs with relatively low levels of foreign operations, (i.e., home-oriented) generate lower rates of return than either those with substantial foreign operations or those with none. An implication is, for FDI to be a strategic motivation for food manufacturing firms, foreign ventures should be operated with intensity similar to that applied at home.

Sources of Strategic Advantage in Foreign Direct Investment

As mentioned earlier, intangible (firm-specific) assets frequent both the theoretical and empirical literature as a source of strategic advantage for firms investing in foreign operations (Casson; Caves; Connor; Grimwade; Grubaugh 1987a and 1987b; Gruber, Mehta and Vernon; Helpman; Meredith). Connor captures the essence of such assets by describing them to include "...patents, trademarks, consumer loyalty to its brands, a positive enterprise image, research and development (R&D) resources yielding technological leadership, effective data gathering and information systems, special relationships with sources of financial capital, and so on" (p. 398). Meredith refers to "marketing management expertise" (p. 111). Grimwade discusses "special marketing skills which enable a firm to differentiate its product from that of rivals..." (p. 167). Substantively, the value of such assets appears to rest on the extent to which firms have developed unique consumer and supplier loyalties; the greater this uniqueness, the higher the value of intangible assets.

Unfortunately, relatively few firms report intangible assets as part of their financial statements; procedures for valuing such assets appear to vary widely among those that do. Among those firms in the panel for which data are available, the intangible asset hypothesis appears to be supported (Table 11.20); MNCs report significantly higher ratios of intangible to total assets than do firms with no foreign affiliates; this relationship holds for both U.S. and non-U.S. firms. Reported differences between U.S. and non-U.S. firms are, again, possibly due to differences in accountancy standards.

To gain additional insight into the sources of strategic advantage in FDI, data on product diversity were compiled. These data include the number of products produced by each firm, classified at the 4-digit SIC level; the number of brand names merchandised by each firm; and the ratio of brand names per SIC. The results are particularly instructive (Table 11.21). Comparing U.S. and non-U.S. MNCs, U.S. firms tend to concentrate on fewer brand names across a larger number of product classes. Comparing foreign-oriented with home-oriented MNCs, similar findings are obtained; foreign-oriented firms merchandise substantially fewer brands across a roughly comparable number of product categories. Consistent with the intangible asset hypothesis, our interpretation is, MNCs with the greatest strategic advantage are those that orient their foreign operations toward intensive merchandising of a limited number of brands; presumably those for which the firm has developed a unique "consumer franchise", complete with whatever idiosyncratic investments and supplier relationships that are essential to such uniqueness. By this measure, foreign-oriented U.S. MNCs appear to have an advantage over their non-U.S. rivals.

TABLE 11.20 Intangible Assets of Leading World Food Manufacturing Firms, 1989/1990[1]

	Intangible Assets as a Percent of Total Assets
Multinational Firms	19.1
U.S. Firms	16.9
Non-U.S. Firms	23.1
Firms Without Foreign Operations	12.4
U.S. Firms	11.6
Non-U.S. Firms	16.7

[1] Sub-sample of 21 firms.

Source: Original data.

TABLE 11.21 Product Diversity of Multinational Food Manufacturing Firms[1]

| | All Firms in Sample | | | Foreign-Oriented Firms | Home-Oriented Firms |
	Average	High	Low		
U.S. Firms					
Number of 4-digit SICs	6.5	14	2	6.8	6.0
Number of Brand Names	29.8	70	6	23.1	39.8
Average Number of Brands/SIC	4.5	14	2	3.4	6.6
Non-U.S. Firms					
Number of 4-digit SICs	5.1	12	2	4.7	5.4
Number of Brand Names	38.2	114	7	21.3	50.9
Average Number of Brands/SIC	7.5	38	3.5	4.5	9.4

[1] Sub-sample of 30 firms.

Source: Original data.

Summary, Conclusions, and Implications

The United States is the world's third largest exporter of manufactured foods and these exports have grown at a real annual rate of 12.6 percent since 1985. Exports from U.S. MNCs have grown even faster, but from a relatively small base. Nevertheless, FDI is far more important as an international market strategy for food manufacturers than is trade.

World-wide, trade in manufactured foods is only one-third as large as sales from foreign affiliates of multinational firms. The difference is greatest for U.S. MNCs. Sales from U.S. foreign affiliates are 9 to 10 times larger than are exports from their U.S. parents. This highlights the importance of using both trade and foreign affiliate sales data (the International Ratio) in accessing the international competitiveness of food manufacturing firms. Trade measures, by themselves, are grossly inadequate as a means of representing the "foreign presence" of most large food manufacturers.

MNCs establish foreign affiliates primarily to access and serve the host country market, rather than to originate exports. Yet, in terms of export propensity, foreign affiliates resemble host country firms more than their parents.

Exports from U.S. firms to their foreign affiliates are more than twice as large as are affiliate exports to the United States. About three-fourths of all U.S. trade with affiliates is intra-firm trade. In general, U.S. affiliates of foreign firms are marginally more trade-oriented than are U.S. parent firms.

Comparing a panel of U.S. and non-U.S. MNCs reveals more similarities than differences. Regardless of nationality, MNCs are larger than firms with no foreign operations; U.S. MNCs, on average, are somewhat larger than non-U.S. MNCs. In all countries, MNCs tend to export a smaller share of their total turnover than do firms with no foreign operations. The international ratio for non-U.S. firms averages twice that for U.S. MNCs, reflecting both higher export and higher FDI propensities. Yet, exports as a share of total international market penetration are the same for both U.S. and non-U.S. MNCs.

Overall, MNCs appear to have lower earnings rates than do non-MNCs. But, those MNCs that originate the largest share of total shipments from foreign operations (foreign- oriented MNCs) have higher earnings rates than either firms with no foreign affiliates or MNCs with relatively low levels of foreign investment. Further, these foreign-oriented firms concentrate their overseas operations on fewer but larger affiliates, product lines, and brand names. Thus, no support was found for the hypothesis, product diversity is positively associated with intensity of FDI among the world's leading food manufacturing firms. Conversely, considerable support was found for the hypothesis, intangible assets are positively associated with such FDI intensity. Both U.S. and non-U.S. MNCs have significantly higher ratios of intangible to total assets than do firms with no foreign direct investment.

Concerning the question, what is the net effect of in-bound and out-bound FDI on trade, the available evidence is mixed. Clearly, this is an area begging additional research. Under what circumstances is FDI trade-diverting, or trade-augmenting? MNCs export a smaller share of their total turnover, on average, than do firms with no foreign operations. MNCs exhibit a preference to produce products in the country in which they are sold for final consumption; this is especially so for branded consumer-ready products. Thus, FDI and trade are to some extent substitute strategies; such a substitute effect appears to be more pronounced for U.S. firms than for non-U.S. food manufacturers.

On the other hand, both U.S. exports of manufactured foods and sales from foreign affiliates have grown rapidly since the mid-1980s. Manufactured food exports by U.S. MNCs grew at an even more rapid pace during this period. Further, U.S. affiliates abroad are a significant market for U.S. exports. On balance, while FDI and trade may be substitute strategies, the net effect of FDI on trade is, most likely, quite small.

Just as trade promotes competition, increases choices in the marketplace, and fosters efficiency, foreign investment seems to promote these same positive effects on economic welfare. Thus, it appears to be important to focus on national regulations that open markets to foreign investment as well as on trade liberalization.

References

Bureau of Economic Analysis. *Foreign Direct Investment in the United States: Operations of U.S. Affiliates of Foreign Companies.* Washington D.C. U.S. Department of Commerce, Economics and Statistics Administration, Bureau of Economic Analysis: Preliminary 1989 Estimates, August 1991.

Bureau of Economic Analysis. *U.S. Direct Investment Abroad.* Washington, D.C. U.S. Department of Commerce, Economics and Statistics Administration, Bureau of Economic Analysis: 1989 Benchmark Survey: Preliminary Results, October, 1991.

Casson, M. "Multinational Firms." In *The Economics of the Firm*, ch. 7, edited by R. Clark and T. McGuinnes. Oxford: Basil Blackwell, 1987.

Caves, R. E. "International Trade, International Investment, and Imperfect Markets." Princeton University, Department of Economics, Special Papers in International Economics, No. 10, November).

Center for Financial Analysis and Research. *Global Company Handbook.* Princeton, N.J.: CIFAR, 1992.

Connor, J. M. "Foreign Investment in the U.S. Food Marketing System." *American Journal of Agricultural Economics* 65(May 1983): 395-404.

Connor, J. M. and D. Neilson Murphy. Unpublished regression analysis. Cited in Connor, J. M. "Determinants of Foreign Direct Investment by Food and Tobacco Manufacturers." North Central Regional Research Project NC-117, Report No. WP-70, March 1983.

De la Torre, J. "Foreign Investment and Export Dependency." *Economic Development and Cultural Change* 23(October 1975): 133-50.

Grimwade, N. *International Trade—New Patterns of Trade, Production and Investment.* London: Routledge, 1989.

Grubaugh, S. G. "Determinants of Direct Foreign Investment." *Review of Economics and Statistics* 69(February 1987): 149-52.

———. "The Process of Direct Foreign Investment." *Southern Economic Journal* 54(October 1987): 351-360.

Gruber, W., D. Mehta, and R. Vernon. "The R & D Factor in International Trade and International Investment of United States Industries." *Journal of Political Economy* 75(1967): 20-37.

Handy, C. R., and J. M. MacDonald. "Multinational Structures and Strategies of U.S. Food Firms." *American Journal of Agricultural Economics* 71(December 1989): 1247-54.

Helpman, E. "A Simple Theory of International Trade with Multinational Corporations." *Journal of Political Economy* 92(1987): 451-72.

Hirschberg, J., J. Dayton, and P. Voros. "Firm Level Data: A Compendium of International Data Sources for the Food Processing Industries." North Central Regional Research Project NC-194, Publication No. OP-34, April 1992.

Horst, T. *At Home Abroad.* Cambridge, MA: Ballinger, 1974.

Horstmann, I., and J. R. Markusen. "Licensing Versus Direct Investment: A Model of Internalization by the Multinational Enterprise." *Canadian Journal of Economics* 20(August 1987): 464-81.

Julius, D. *Global Companies and Public Policy.* New York: Council on Foreign Relations Press, 1990.

Kravis, I. B. and R. E. Lipsey. "Sources of Competitiveness of the United States and of its Multinational Firms." *Review of Economics and Statistics* 74(May 1992): 193-201.

Lipsey, R. E. "Foreign Direct Investment in the U.S. and U.S. Trade." National Bureau of Economic Research Working Paper No. 3623, February 1991.

Meredith, L. "U.S. Multinational Investment in Canadian Manufacturing Industries." *Review of Economics and Statistics* 66(February): 111-19.

Organization for Economic Cooperation and Development. *Industrial Structure Statistics—1988*. Paris: OECD, 1991.

Porter, M. E. *The Competitive Advantage of Nations*. New York: The Free Press, 1990.

Reed, M. R. and M. A. Marchant. 1992. "The Global Competitiveness of the U.S. Food-Processing Industry." *Northeastern Journal of Agricultural and Resource Economics* 21(April 1992): 60-70.

United Nations. "Statistical Papers, Commodity Trade Statistics, According to Standard International Trade Classification, Series D." Statistical Office, Department of Economic and Social Affairs (magnetic tape), 1990.

12

Assessing the Competitiveness of the Canadian Food Sector

William M. Miner

The Canadian food industry, like many other sectors of the economy, is being challenged to compete in the rapidly changing global environment of the 1990s. Faced with strong pressures for change from within and outside the sector, and new opportunities in the continental market and overseas, Canadian food processors are adjusting and restructuring in response to the reality of a more open and competitive environment.

The forces of change are diverse and compelling. They include increased flows over borders of capital, goods, services, technology, and corporate equity. Flexible exchange rates, freer trade, rapid communications, deregulation, and improved distribution systems are powerful catalysts in this process. In the food sector, the directions of change are dictated primarily by evolving living patterns and consumer diets that demand further processed, differentiated and packaged foods. There is an increased emphasis on convenience, nutrition, food safety and environmental considerations. At both the production and processing levels, technological improvements, economies of scale and better farming and processing methods have led to an amazing improvement in productivity. These developments not only accelerate the pace of change and increase competition, but they also place strong pressures on the policy environment. Traditional farm and trade policies and programs have become too expensive to sustain, and many are no longer effective in meeting their objectives. Further, they are causing serious market distortions and higher costs. It is now generally recognized that certain policies and regulations weaken the ability of the farm and food sector to compete.

In Canada, these developments have stimulated an intense interest in assessing the competitiveness of the food industry. As part of a general Agricultural Policy Review initiated by Agriculture Canada, a task force led to the establishment of an Agri-Food Competitiveness Council. Organized as a private, nonprofit advisory group, the Council's stated objectives are listed in its "Prosperity Initiative"

- to create a climate for change,
- to encourage strategic integration in the food chain,
- to encourage innovation,
- to create a favorable economic environment that will encourage investment in the sector,
- to enhance human resources,
- to assist the sector in creating consumer value, and
- to encourage environmental sustainability.

In a related initiative undertaken by the Food Policy Task Force of Industry, Science, and Technology Canada (ISTC), a wide-ranging program of studies was organized to examine the factors contributing to competitiveness and to assess the relative position of food and beverage processors. Its purpose was to assist food processing companies in meeting the unprecedented challenges and opportunities of the rapidly changing competitive environment. In addition, ISTC released an assessment of the major industrial sectors, including food and beverages, as part of a general "Prosperity Initiative" to promote international competitiveness (ISTC 1991). Although some of the studies were preliminary and not all have been published, a review of this work provides an indication of the competitiveness of Canada's food industry (Miner, 1991). The studies supported extensive consultations with the private sector in a project entitled "Getting Ready to Go Global" (ISTC 1992).

The Industry in Perspective

Canada's food industry developed from a relatively strong and extensive resource base with growing and handling conditions that were conducive to high volume, quality production of most temperate agricultural commodities. There are significant regional differences in agriculture (the main export crops and livestock dominate production in the Western provinces whereas agriculture is more diversified in other regions), but food processing is undertaken in all parts of the country based on local raw materials. However, the further-processed food and beverage industries are concentrated near the larger urban areas, particularly in Ontario and Quebec.

The grains, oilseeds, and red meat subsectors developed on an export basis and continue to be relatively competitive in world markets. This was also the case for a range of other farm products including pulses, seeds, selected fruits and vegetables, breeding livestock, and maple syrup. Some milk products were export competitive in the past but the dairy and poultry subsectors are now highly managed and protected. Temperate fruit and vegetable production had relatively high tariffs, particularly on a seasonal basis, which are now being reduced.

The Canadian food processing industry was established to serve a limited and protected domestic market. Initially comprised of small- and medium-sized Canadian firms and branch plants of larger foreign corporations, its exports were primarily to the nearby U.S. market or Commonwealth countries under preferential tariff arrangements. As tariff levels declined, a gradual process of rationalization has moved the industry toward a more diversified and differentiated product sector. Trade is becoming more important to the processed food industry although there are wide differences among product groups. Meat processing, vegetable oils, some frozen fruits and vegetables, and distillery and cereal products are export-oriented and relatively competitive. The fruit, vegetable, wine, and confectionary product subsectors have greater import penetration and appear to be less competitive. The dairy and poultry product subsectors, which operate under supply management, are not price competitive.

The United States remains Canada's principal foreign market for processed foods and beverages, and this trade continues to expand. Japan has replaced Britain and other European Community countries as the second most important export market—and Canadian food products are moving to a growing list of destinations.

Taken together, the primary agriculture and food industries, including wholesale, service, and retail activities, contribute about 8 percent of Canada's gross domestic product (GDP) and employ 673,000. These industries generate a substantial trade surplus of at least $2 billion annually compared to recent exports exceeding $10 billion. Approximately one-half of farm cash income is earned from exports. Canada has maintained a 3.5-4.0 percent share of world agri-food exports in recent decades. The food and beverage processing industries provide about 2.4 percent of GDP and represent Canada's second largest manufacturing sector. Exports represent about 10 percent of the value of shipments of processed food and beverages (excluding fish products). The composition of the food and beverage sector and its share of value-added in the domestic market is summarized in Table 12.1.

Canada traditionally relied more on market intervention than government expenditures to support the farm sector, given its relatively small population and tax base. A split jurisdiction for agriculture between federal and provincial governments has broadened the range and complexity of the policies and structures that influence the agri-food industry. Policies are a blend of export-oriented programs to enhance volume shipments of bulk and partially pro-cessed commodities and to protect and stabilize production for the domestic market. Intervention mechanisms include subsidized transportation rates and central desk selling for cereals, producer marketing boards for a range of crops and livestock, and supply management and import quotas for dairy and

TABLE 12.1 A Profile of Canada's Food and Beverage Industries, 1989

Industry Name	No. of Establishments	Total Shipments ($000,000)	Total Value Added ($000,000)	Total Employed
Meat and Fish Products	1,101	15,681	3,366	76,123
Processed Fruits and Vegetables	227	3,719	1,560	18,169
Dairy Products	372	8,700	2,364	25,920
Grain Products	600	5,251	1,376	16,160
Vegetable Oil Mills (except corn oil)	11	1,017	127	1,112
Bakery Products	529	3,318	1,641	29,042
Sugar and Confectionery Products	118	1,997	809	10,916
Other Food Products	427	5,473	2,405	24,664
Food Industries	3,385	45,158	13,648	202,106
Soft Drink Industry	150	2,351	872	8,457
Alcoholic Beverages	124	4,122	2,751	19,271
Beverage Industries	274	6,473	3,623	27,728
Food and Beverage Industries	3,659	51,631	17,271	229,834
All Manufacturing Industries	39,150	356,636	135,636	1,970,259

Source: Statistics Canada, "Survey of Manufactures, 1989."

poultry. Safety net programs exist for most commodity subsectors and are gradually shifting toward income rather than market support. The red meat subsector has remained relatively free from market intervention programs. Tariffs are most significant for the fruit, vegetables, and processed food subsectors. Provincial policies aimed at increasing self-sufficiency and employment have the greatest impact on the marketing of wine, beer, spirits, dairy, and poultry products and some fruits and vegetables. Grading, health, and technical regulations are comparable to or exceed international standards.

Agricultural support policies, both federal and provincial, are now changing in response to global market integration, fiscal pressures, and new trading arrangements. The Free Trade Agreement with the United States is forcing the pace of industry rationalization that has been underway for some time in the food and beverage industry. Increased competition, expenditure restraints, deregulation, and new corporate affiliations are expected to lead to lower subsidies, the gradual removal of border restrictions, and less intervention in the domestic market.

Analyzing Competitiveness — A General Review

The studies reviewed in relation to the work of the ISTC Food Policy Task Force and the Agri-Food Competitiveness Council defined competitiveness as the sustained ability to profitably gain and maintain share in domestic and export markets. This definition appears to be most appropriate for an individual firm or an industry subsector but may be misleading when used for entire industries comprised of many firms with different structures and operations. It is necessary to take into account the context in which the industry is competing. Macroeconomic policies, trade policies, and government interventions in the market may influence factor costs and profits. Furthermore, the global environment and domestic policies are changing; hence assessments based on past performance may not be indicative of current or future competitiveness.

The Canadian program of studies noted earlier used a number of indicators of profitability and performance as a basis to analyze competitiveness. These included market share, export orientation, import penetration, productivity of labor or other factors, application of technology, range and quality of products, relative input costs, product differentiation, innovation, demand conditions, and growth in sales or profits. This wide range of factors provided a general indication of Canadian food industry competitiveness, but the information was considered to be inadequate to draw definitive conclusions. Some studies looked in greater depth at certain subsectors, permitting a more careful examination of the relative performance of four subsectors: bakery and other wheat-based products, dairy products, processed fruits and vegetables and poultry products. Since much of the analysis was aimed at assisting the industry to adapt to a more open continental market, the comparisons usually related the Canadian performance to the U.S. food industry. This work is supplemented by a recent review of changes in the performance and structure of Canada's food and beverage processing industries prepared by Agriculture Canada (Barkman, 1992). Some general observations can be made on the competitiveness of Canada's food industries based on these reviews.

The value of Canadian output of processed foods and beverages grew quite quickly during the 1970s, but the rate of growth tended to plateau in the last decade. Output in the United States has continued to increase. Employment in the Canadian sector declined during the 1980s as in the United States. However, at the turn of the decade, this trend appears to have reversed in the United States whereas the decline continued in Canada.

The growth in labor productivity in the Canadian food sector was much slower than in the United States during this period. Multifactor productivity growth was also somewhat slower in the Canadian than the U.S. food processing industry. In both countries, the number of food processing establishments declined by 20-25 percent over the last two decades. But the downward trend slowed in Canada in the 1980s compared to U.S. experience. Apparently the U.S. industry is rationalizing its operations more rapidly, which is reflected in higher productivity measurements.

The average size of operations in terms of value-added and employment has increased in both countries, but U.S. companies are significantly larger on average than Canadian companies. However, the relatively smaller Canadian firms may have advantages in exploiting niche markets through greater flexibility and selective specialization.

Several studies indicate that research and development in the Canadian food industry is low in comparison with the United States. Most technology applied in the sector is imported. It is recognized that U.S. firms have advantages in R&D, technology transfer, and systems innovation due to the size and scope of their market. However, new plant investment in Canada utilizes the latest techniques and systems, and hence the restructuring process should close the gap between the two industries. Product and marketing innovations were generally viewed as lower in the more protected food subsectors in Canada.

Trade performance provides an indication of competitiveness, although allowances must be made for products not grown in Canada and the significance of trade barriers and subsidies. Overall the Canadian food industry is more oriented toward export markets than its U.S. counterpart. The Canadian food and beverage industries have seen their export orientation and import penetration increase steadily since 1970. The Canadian agri-food industry maintained its relative export position and held its own against import penetration during the 1980s. Exports have moved from 10 percent of production in 1970 to 15 percent in 1988, while imports have moved from 7 percent of production in 1970 to 12 percent in 1988. But these industries continue to rely heavily on low value-added exports. As indicated in Table 12.2, some subsectors have improved their export orientation since 1981, particularly red meats, some cereal products and distillery products. The trade experience for other categories is mixed, although exports are important for oilseed products, and some fruits and vegetables. It is difficult to assess the international competitiveness of Canadian food products as markets are severely distorted by export subsidies.

It is evident from the studies that some raw material costs are higher-priced in Canada than the United States. This was most obvious for poultry and dairy products, but also the case for wheat-based products and some fruits and vegetables. There were other cost disadvantages identified, including oil-based energy, pesticides, and transport and distribution charges. A few commodity inputs were less costly, such as sugar. Some utility costs were lower including electricity and natural gas. Net labor costs were found to be generally lower in Canada but significant regional and sectoral differences were identified. Food packaging costs were considered to be higher in Canada, but the disparity is narrowing as a result of tariff reductions.

TABLE 12.2 Export Orientation and Import Penetration in Canada's Food and Beverage Industries, 1981 and 1987 (percent)

Processors and Manufacturers	Export Orientation		Import Penetration	
	1981	1987	1981	1987
Meat and Fish Products	27.8	38.5	17.8	13.2
Processed Fruits and Vegetables	9.3	8.5	27.7	23.3
Dairy Products	4.3	2.1	2.0	2.0
Grain Products	13.1	8.3	2.7	4.5
Vegetable Oil	24.4	25.4	22.5	27.9
Bakery Products	3.7	7.8	3.0	5.2
Sugar and Confectionery Products	12.8	32.1	48.5	58.2
Other Food Products	3.8	5.0	24.7	22.3
All Food Products	12.9	14.5	12.4	12.3
Soft Drink Industry	0.6	0.8	2.3	3.1
All Beverages	12.6	9.8	10.0	9.1

Source: Industry, Science, and Technology Canada, "Commodity Trade by Industrial Sector," 1981 and 1987.

Although most industry sectors in Canada exhibited continued growth in value-added and positive trends in productivity, these indicators trailed similar measurements in the United States and the gap appears to be widening. For example, the upward trend in higher value-added products as a share of sales continued in both countries during the 1980s for wheat-based products, processed fruit and vegetables, and poultry products; however, Canada is falling further behind the United States. A similar picture emerges from labor productivity, measured by value-added per product worker.

There is considerable evidence that the productivity of the food and beverage industries has been compromised by federal and provincial policies designed to protect markets and manage supplies. For a number of products, the national market is fragmented because of these protected markets. Many of the studies under review pointed to public policies that have created an environment in Canada that makes it more difficult for food processing firms to compete. The use of market intervention mechanisms and border controls to provide farm income support results in a higher cost structure and weakens the ability of the sector to reduce costs. There appears to be a link between a high level of market management with border protection and a lack of competitiveness. These policies tend to discourage innovation and industry rationalization and inhibit the development of higher value products.

An Overview Assessment of Competitiveness

The following assessment emerges from this review of the competitiveness of the Canadian food processing industry.

- Despite data limitations and uncertainty over the relationship between value-added indicators and industry performance, there are many indications that the Canadian food processing industry is, in most cases, less competitive than its U.S. counterparts.
- The gap in performance appears to be widening but adjustments are under way and more analysis is needed.
- The least competitive subsectors are the most highly regulated.
- The emergence of integrated, global markets and free trade in North America is adding urgency to the need to alter policies that discourage value-added activities and subsector competitiveness.
- The Canadian agri-food sector does appear to be holding its relative position in export and import penetration, but it continues to rely on lower-valued products.
- The food and beverage processing industries are adjusting rapidly and governments must move quickly to ensure that the policy environment encourages investments and value-added activity in Canada from the restructuring process.
- Policy and operational adjustments to encourage industry competitiveness should focus on:
  - providing access to competitive raw materials and other inputs within a realistic time-frame through changes in supply management and other intervention systems in addition to the relaxation of border restrictions,
  - the removal of interprovincial barriers to trade to provide a domestic base for industry rationalization,
  - steps to ensure that tax structures, antitrust legislation, and regulatory and inspection services are conducive to developing a competitive food processing industry,
  - an environment developed in Canada to foster private innovation, research and development and human resources, including improved education, training and development programs, and
  - the development of intercorporate relationships, marketing strategies, and supplier/customer linkages to help develop competitive subsectors.

Despite the analysis already done, which is continuing and improving, the issue of competitiveness may be even more complex. The global factors influencing competitiveness are changing rapidly, and studies of relatively recent developments may not capture the essence of future competitiveness.

The comparison of relative costs between countries does little to reveal future competitiveness. In the integrated global market, where there are relatively few growth sectors, the development of competitive industries is also related to savings and investments. Growth industries are usually perceived as those based on using your heads rather than your hands. These require educated, innovative, entrepreneurial people and a positive climate for investment. Examples of growth industries include microelectronics, biotechnology, telecommunications, and computers. It is instructive for Canadians to note that these do not depend on traditional advantages in natural resources, transportation or handling capabilities. Although Canada should not ignore its resource-based sectors in developing its competitive industries, it appears that the human resource, particularly its innovative and organizational skills, represents the basic ingredient for success. However, the issues of education, innovation, and creating an entrepreneurial environment require long-term commitments by both industry and government and will take much time to yield dividends.

It is important to recognize that agriculture in all countries is being shaped by strong influences from within and outside the sector. Markets are becoming integrated and competitive on a regional and global basis. These developments are forcing farm groups, businesses, and governments to re-examine their policies and programs. It is possible to observe trends and to identify influences that will have a bearing on future competitiveness, such as value-added activity. Many of these trends that are forcing adjustments in policies and industry structures appear to be inevitable almost regardless of the activities of governments. But policymakers must respond to the present situation and these responses will be more appropriate if they are based on sound research and analysis.

The Canadian agri-food industry enjoys certain advantages that can be exploited in a favorable economic and policy environment. As indicated, government policies and corporate structures and operations are under intensive review in Canada, and considerable policy adjustment and industry reorientation is under way. Farm income support programs are being decoupled from individual commodities in the export-oriented subsectors and most programs are being adjusted to increase their sensitivity to market developments. Considerable rationalization is under way in the primary processing industries and new corporate affiliations with established multinational enterprises have recently emerged. The further processed food subsectors are consolidating and adjusting their operations on a continental basis.

References

Agri-Food Competitiveness Council. "Prosperity Initiative: Agri-Food Sectoral Response." June, 1992.

Barkman, P. "Overview of Changes in the Performance of Canada's Food and Beverage Processing Industry." Agriculture Canada Policy Branch, Working Paper, No. APD 92-3, July 1992.

Industry, Science and Technology Canada. "Industrial Competitiveness: A Sectoral Perspective." Consultation Paper, 1991.

————. "Getting Ready to Go Global: Summary Report of Regional Consultations with the Canadian Food and Beverage Processing Industry." June 1992.

Miner, W. M. "Competitiveness and the Canadian Food Processing Industry:An Overview — ISTC Program of Studies." Food Policy Task Force Working Paper June 1991.

13

Market Mass Competitiveness in the Canadian Food Industry

Tim Hazledine[1]

Introduction

This chapter develops a new measure of competitiveness called "market mass" and uses it to examine the competitive position of forty Canadian food and beverage manufacturing subsectors relative to their U.S. counterparts. The year of comparison is 1986, which is just before the implementation of the Canada/U.S. free trade agreement (CUSTA). The novelty of the market mass measure is not fundamental; on the contrary, it works with what, in Canada, has become a fairly standard definition of competitiveness. The contribution is to offer a way of implementing empirically this definition, which differs somewhat from those previously proposed.

One purpose of the chapter, then, is to describe the market mass index and estimate it for forty Canadian agricultural subsectors. The second purpose is to explore some of the properties of this new measure. Is competitiveness related to productivity, trade protection, or market concentration? These and a number of related questions will be addressed below.

Measuring Competitiveness as Market Mass

Competitiveness is a comparative concept. A firm or industry or country is measured as being competitive relative to another such entity. The entities to be compared here are Canadian and U.S. industries using the following definition:

> The index of competitive performance for a Canadian industry is its output relative to the output of its U.S. counterpart, were it to charge the same price as the U.S. industry.

One can find a benchmark for this measure in the relative sizes of the two economies. Real Canadian GNP is about 9 percent of the United States, so that would expect an "average" Canadian subsector to produce about 9 percent as much as its U.S. counterpart. Therefore, one can say that Canadian subsectors that would succeed in selling more than 9 percent of U.S. output, were they to charge the same price as their American competitors, are "more competitive" than the counterpart American subsector.[2]

It is also natural to rank the Canadian subsectors against each other, by imputing superior competitiveness to those subsectors with higher values of the index. This is done below, but a conceptual caveat is noted: Such a comparison is strictly valid only if differences in index values are due only to differences in Canadian subsectors' performance — that is, not to variations in the efficiency of U.S. food subsectors. However, in almost all situations, it is only competition from the United States that is relevant for Canadian firms, so the ranking is likely to be useful from a practical viewpoint.

Now, compare the definition above with that proposed by the Canadian Task Force on Competitiveness in the Agri-Food industry, and adopted by van Duren, Martin and Westgren (1992) and Ash and Brink (1992).

> Competitiveness is the sustained ability to profitably gain and maintain market share.

It will be argued that this index is the appropriate empirical implementation of the concept underlying the task force definition. Thus, this chapter differs from van Duren et al., to whom the definition points to a multidimensional index. In particular, they single out profitability and market share as separately measurable competitiveness indicators. There is certainly precedent for this interpretation, but it is inappropriate in the context of food processing subsectors, for three reasons.

First, a multidimensional measure is likely to lead to ambiguities. Is an industry with high profits and low market share more or less competitive than one with low profits and high share? Clearly, a unidimensional measure is to be favored in principle, so long as it can capture the essence of competitiveness.

Second, profitability itself is an unreliable normative metric. In mature manufacturing, technologies are well-known and inputs readily available. In these circumstances, free competition should tend to force rates of return to equality across subsectors. Industrial Organization (IO) specialists therefore attribute persistent differences across industries in profitability as being due to differences in market power so that higher profits are expected to have negative effects on economic efficiency.[3]

Against this, it should be noted that, even in mature processing industries, there may be difficult-to-replicate sources of efficiency differences (i.e., rent-yielding assets) that are rewarded with higher profit margins and that may certainly be of legitimate interest to the student of competitiveness. But how are these efficiency rents to be disentangled from market power rents?[1]

Therefore, the word "profitably" in the task force's definition is a constraint, not an independent objective. Profitability must be adequate to sustain the activities of the firm, subsector, or industry, and "competitiveness" must not be "bought" by means of artificially low profitability. This constraint should not be of concern to food processing, which like most manufacturing activities operates in an environment in which the private capital markets can be relied on to weed out subnormal profit performance, and gets little direct government assistance. It should be noted that primary industries may differ — one thinks of the vast subsidies being poured into national wheat subsectors to maintain global market share.

The third point follows from the second and concerns the validity of actual output or market share as another dimension of competitiveness. According to IO theory and evidence, the major instrument whereby market power is translated into monopoly profits is noncompetitive pricing (Connor and Peterson, 1992 and Connor et al., 1985). With subsector demand being typically quite inelastic, oligopolistic firms can increase profitability by pushing up the market price. Indeed, the ability to raise price above marginal cost is a standard measure of market power. That is, oligopolists can tradeoff output or market share for higher prices or profit margins; thus, differences in actual share can reflect differences in market power, not in fundamental competitiveness.

The measure proposed here avoids these difficulties because it is based on the position of the subsector's demand curve, not on the position the subsector chooses on its curve. That is, it is the ability to "gain and maintain market share," not the actual share, that matters. Figure 13.1 illustrates these points. The ratio of Canada to U.S. output price is on the y-axis for each subsector. The ratio of real outputs of the two countries is on the x-axis. If the actual relative price and output are above and to the right of A — say, at point B — then the Canadian subsector is clearly more competitive than the U.S., since there is no way that a nonpathological demand curve that passes through B can find its way below and to the left of A. Similarly, a price/output coordinate at point D would imply an unambiguously less competitive Canadian subsector.

Points to the northwest and southeast of A are the interesting cases where a bidimensional price or profit-plus-actual market share criterion is uninformative. One doesn't know the underlying competitive position without knowing more about the subsector's demand curve. That is why Canadian prices are adjusted to U.S. prices for comparison. Taking both price and output in the U.S. as exogenous to Canada, one can plot an adjusted demand curve for the Canadian subsector in terms of relative price and output. Then, if this adjusted, demand curve goes through point A on Figure 13.1, the Canadian subsector is said to be equally competitive with its U.S. counterpart.

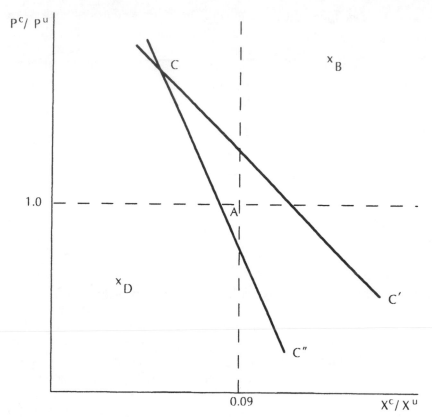

FIGURE 13.1 Measuring Competitiveness.

Existing econometric estimates of subsector demand elasticities are used to calculate where an adjusted demand curve going through observed points crosses the horizontal line representing price parity (Pc/Pu =1). The corresponding value of relative output is then the measure of competitiveness, as defined at the top of this section.

This competitiveness index can be interpreted as reflecting the subsector's ability to generate revenue (to enhance customers' willingness to pay) in its output markets. It is a measure of the size or mass of a subsector's market, and will be here dubbed "market mass."

The measurements of market mass reported below should be of intrinsic interest, since it is an attempt to implement empirically a fairly widely accepted definition of competitiveness. However, the usefulness of the measure for policy purposes depends on an ability to explain it. What factors influence success in achieving high market mass?

Several papers in the present volume assert the importance of productivity as a key determinant or "driver" (van Duren et al.'s word) of competitiveness.

Lau (1992) and Kalaitzandonakes, Gehrke, and Bredahl (1992) appear to go a step further and use productivity as their measure of competitiveness performance itself. Further, in Canada, concern with allegedly poor domestic productivity in manufacturing (compared with the United States) has been very influential in determining the course of trade policy. Since the 1960s, it has been argued that Canadian tariff structure has fostered a high-price/high-cost industrial structure (Eastman and Stykolt, 1967 and Wonnacott and Wonnacott, 1967). Under the umbrella of protection, domestic oligopolists set high prices, which encouraged entry and overcrowded domestic industries. This prevented firms from achieving efficient production scale and raised unit costs. Thus, all the tariff rents were dissipated in cost inefficiency. The policy implication, now realized in the CUSTA that began coming into force in 1988, was obvious: free trade. In the simulations of the general equilibrium model of Harris and Cox (1984), free trade-induced productivity improvements were predicted to result in striking increases in the export performance of Canadian manufacturing industries.

How does productivity fit in conceptually with market mass? If the product market is fully integrated and supplied by firms producing a standardized product, as is assumed in both the standard textbook competitive model and in the Canadian model set out above, then one can make two predictions. First, since arbitrage or oligopolistic collusion enforces price parity between suppliers, market mass will be the same as actual market share. Second, this market share will be entirely determined by relative productivity, assuming that inputs are freely traded. If there are nontradable inputs, then performance will depend on relative costs (i.e., relative productivity times relative input prices).

If markets are not fully integrated, as they were not in North America before the CUSTA, then price parity doesn't exist and market mass can differ from actual share. Productivity will still be the major driver of competitiveness and it will be inversely related to tariff protection through relative plant sizes in Canada and the U.S.

Is there an alternative to the productivity model of competitiveness? Yes there is, though it is not nearly so clearly worked out analytically. It starts by dropping the standardized product assumption. Instead, it is assumed that, even in subsectors producing physically quite homogeneous commodities, actual transactions are invariably heterogeneous. Transactions differ in delivery terms, payment arrangements, risk of quality problems, size, etc. In consumer goods industries, brand image, which may or may not reflect "objective" differences in products, is a major factor. All this adds up to firms or groups of firms "owning" their own demand curves, and success being measured in the position of those demand curves, which will be picked up at the national level by the market mass index.

What will be the determinants of success in this differentiated-transactions environment? What determines firms' (or countries') ability to increase customers' willingness to pay by offering attractive design, quality, financial terms, brand image, etc.? If I could answer this I would be one of those very rich business school gurus, rather than a humble economics professor. However, the basic ability to transform physical inputs into physical outputs (productivity) must be expected to recede into a minor role. In any case, this will be tested below. A variety of "industrial organization" variables (for example, market concentration, foreign ownership, and tariff protection) that could plausibly affect market mass are analyzed later.

Data

The main data sources are the 1986 Canadian Census of Manufactures, the 1987 U.S. Census and 1986 U.S. Survey of Manufactures. The unit observed is the subsector, defined by the U.S. 4-digit Standard Industrial Classification (SIC). The Canadian SIC is often more aggregated than the U.S., so Canadian subsector data were disaggregated to fit the U.S. classification using U.S. relative factor proportions as a guide, but requiring that the constructed numbers add up to the known Canadian SIC aggregate.

The result is a database with 40 subsectors — sixteen more than would have been available had the sample been restricted to the Canadian 4-digit SIC. The Censuses provide information on value of shipments, expenditures on materials, energy and wages/salaries, plant size, and, for the U.S., capital stocks and concentration ratios. Canadian concentration ratios and data on foreign ownership were disaggregated from the Canadian SIC using a variety of extraneous information. Data on advertising intensities and on the number of midsized plants in the U.S. were from Connor et al.

Estimates of price elasticities of demand are from Pagoulatos and Sorenson's (1986) econometric study of the U.S. market for manufactured food and beverage products. These estimates play an important role in the competitiveness calculations, as they are the means to obtain the relative outputs predicted if Canadian subsectors charged the same price as their U.S. counterparts. There are at least three potential problems with the Pagoulatos and Sorenson elasticities. First, they are subject to estimation error (in particular, they may be biased towards zero when there are errors in data or insufficient variation in prices). Second, they reflect the behavior of U.S. American, not Canadian, consumers — however, incomes and tastes in the two countries are similar enough to make this a minor source of error.

Third, elasticity is defined for the situation of all own-prices moving together, whereas the counterfactual used to calculate market mass has Canadian prices changing to parity with (unchanged) U.S. prices. If they were in fact competing in an open North American market, then the relatively small Canadian subsectors could expect a much larger demand response to such a price change than is predicted by the total-market elasticity. For this reason, these elasticity estimates certainly should not be used to predict the effect of price changes on demand for Canadian output under free trade. However, as shown below, in the pre-free trade environment of 1986, tariffs appear to have been high enough to virtually segment the North American market — very little cross-border trade actually took place. Therefore, subject to the caveats from the first and second problems noted above, it should be valid to use these elasticities to calculate pre-free trade market mass.

The data that really make the study possible (and unique) are the measures of relative Canada/U.S. output and input prices for each subsector. The output prices were constructed from comparisons of unit values for as many commodities as possible for which the Canadian and U.S. definitions appear to be reasonably similar. When more than one commodity's relative unit value or price was measured in a particular subsector, unit values were aggregated using a Tornqvist (log-linear) index with average Canada-U.S. shipment shares as weights. A total of 109 output prices were calculated, covering 81 percent and 65 percent, respectively, of the shipments of the Canadian and U.S. subsectors.

Input prices were constructed similarly. Tornqvist indices summarize Canada/U.S. differences in "variable" input prices (materials, energy, and labor). Generally, prices for material and energy inputs are simpler to measure than the outputs into which they are converted, and it was not difficult to obtain reliable data on relative prices paid for materials, which are by far the largest component of input costs in the food sector. Labor costs presented more problems (only actual wages and salaries per employee were available) so that comparisons reflect only differences in true costs if differentials do not embody quality premia.

The relative output prices are used to deflate values of shipments in order to estimate relative real outputs. Thus, real output measures have at least two sources of error: from measurement errors in the price indices and from the omission of prices of commodities produced by the subsector but not included in the price index. In any case, with data on relative prices, relative real outputs, and the estimates of demand elasticities, it is possible to compute the measure of market mass competitiveness for each subsector.

Trade flow data are built from commodity-level databases of Statistics Canada. These data refer only to trade in the commodities for which relative prices were available.

Results

This section investigates 24 variables set out on four tables (variable definitions are given in the Appendix). The market mass competitiveness index is shown on column 1 of Table 13.1. It shows the ratio of Canadian to U.S. subsector real output if both subsectors charged the same price. The table ranks the subsectors from highest to lowest market mass and also shows mean values

TABLE 13.1 Market Mass Indicies and Associated Variables for Forty Canadian Industries

	1	2	3	4	5	6	7	8
	CDA/US Market Mass Compet. Index	CDA/US Output Price	CDA/US Real Output	CDA/US Input Prod.	CDA/US Resource Costs	Canada Price/Cost	CDA/US Price/Cost	CAN/US Unit Value (1988)
Butter	0.283	1.13	0.270	1.03	1.24	0.98	0.94	1.38
Malt	0.246	0.97	0.248	0.95	0.87	1.38	1.05	1.51
Feed	0.174	0.93	0.175	1.05	1.06	1.10	0.93	1.14
Pork	0.159	0.96	0.164	0.98	1.00	1.04	0.93	1.10
Chewing Gum	0.145	1.05	0.144	0.64	0.88	1.72	0.76	0.80
Peanut Butter	0.138	0.71	0.145	0.97	1.08	0.95	0.63	0.79
Juice	0.137	0.85	0.142	0.95	1.06	1.09	0.73	1.79
Processed Pork	0.135	1.01	0.134	0.96	0.98	1.14	0.99	1.39
Inedible Tallow	0.122	1.05	0.119	0.91	1.00	1.35	0.95	0.64
Canned Vegetables	0.119	1.32	0.113	0.87	1.01	1.72	1.13	1.74
Canned Mushrooms	0.118	0.87	0.122	0.93	0.86	1.28	0.95	1.68
Sugar	0.118	0.60	0.127	0.87	0.49	1.32	1.08	0.85
Beer	0.111	1.17	0.106	0.96	0.99	1.85	1.13	1.46
Flour	0.104	1.44	0.100	0.95	1.52	1.09	0.90	1.30
Frozen Vegetables	0.102	0.91	0.104	0.80	0.74	1.56	0.98	1.09
Cake Mix	0.099	0.76	0.100	1.14	1.02	1.29	0.84	0.60
Cheese	0.098	1.30	0.083	0.99	1.22	1.23	1.06	1.43
Pasta	0.095	0.96	0.096	0.77	1.24	1.13	0.60	0.87
Beef	0.095	0.99	0.095	1.00	1.01	1.08	0.98	1.11
Ice Cream	0.094	1.26	0.087	0.91	1.00	1.43	1.15	1.52
Bread	0.084	0.86	0.086	1.04	1.17	1.29	0.76	0.83
Milk	0.082	1.43	0.077	0.78	1.16	1.19	0.97	1.44
Tea	0.080	0.92	0.081	0.89	1.00	1.71	0.82	0.84
Poultry	0.079	1.26	0.070	0.83	1.10	1.14	0.94	1.17

Milk Powder	0.078	1.09	0.079	0.96	1.22	1.14	0.86	1.35
Smkng Tobacco	0.077	1.07	0.076	0.93	0.94	2.76	1.06	1.15
Jelly & Jam	0.075	1.27	0.071	0.90	1.39	1.32	0.82	1.16
Starch	0.073	1.25	0.071	0.92	1.13	1.41	1.01	1.72
Coffee	0.072	1.10	0.071	1.17	1.35	1.32	0.95	1.10
Sausages	0.068	0.93	0.072	0.95	0.96	1.15	0.92	1.20
Chip & Popcorn	0.067	1.21	0.066	0.60	0.96	1.57	0.76	1.30
Breakfast Cereal	0.063	0.63	0.064	1.05	0.90	2.02	0.74	0.58
Vegetable Oil	0.061	1.05	0.060	1.08	1.15	1.06	0.98	0.84
Biscuit	0.056	1.10	0.055	0.62	1.00	1.32	0.69	1.88
Shortening & Margarine	0.054	0.95	0.055	0.97	1.08	1.08	0.85	0.92
Confectionery	0.053	0.94	0.054	0.90	0.93	1.45	0.91	1.05
Soft Drinks	0.052	1.01	0.052	0.89	1.02	1.24	0.83	1.06
Peanuts	0.050	0.85	0.051	0.94	1.07	1.11	0.75	0.78
Dog & Cat Feed	0.044	1.01	0.044	1.01	1.06	1.69	0.86	0.79
Wine	0.040	1.62	0.037	0.80	1.22	1.51	1.06	1.91
Mean Top-10	0.166	1.00	0.165	0.93	1.02	1.25	0.90	1.23
Mean Bottom-10	0.054	1.04	0.054	0.89	1.04	1.41	0.84	1.11
Mean All	0.100	1.04	0.099	0.92	1.05	1.36	0.91	1.18
Significance of Mean Differences	**	—	**	—	—	—	—	—

for the top- and bottom-10 subsectors, as well as for the whole sample of forty. In addition, the table also ranks the results of a t-test on the significance of the difference in means between top- and bottom-10 subsectors.

There is a wide range of competitiveness among the forty food and beverage subsectors. The top subsector is somewhat of a surprise (butter), but surely no one who has sampled the Canadian-made product will be surprised by the identity of the least competitive subsector (wine). On average, output ratios at price parity are predicted to be more than three times larger for the top-10 than for the bottom-10 group.

On average, the market mass of the forty Canadian subsectors is just above the "normal" 9 percent of U.S. output. This figure should not be used to make inferences about the overall competitive position of the Canadian food and beverage sector, because the sample is biased towards the products and subsectors in which Canada specializes. There are several U.S. 4-digit SIC food subsectors that are not included at all in the study simply because they are not represented at significant levels in the Canadian industrial structure. The forty subsectors account for 87 percent of total food and beverage shipments in Canada, and rather less (82 percent) in the U.S. Obviously, the U.S. is much more competitive than Canada in these omitted activities.

Other variables in Table 13.1 represent just about all the commonly cited "drivers" of or proxies for competitiveness. Column 2 gives the actual relative Canada/U.S. output price. On average, Canadian prices are 4 percent higher than U.S. prices in 1986, but given the large intragroup variations in prices, there is no significant difference between the groups. That is, there is no systematic difference in pricing strategies between the most and least competitive subsectors.

Column 3 gives the actual ratio of Canadian to U.S. real output. There is in fact very little difference between these numbers and the market mass numbers of column 1. This is not because price ratios are generally close to parity, but because the low estimates of price elasticities of demand mean that even quite large movements up or down adjusted demand curves do not generate large differences in quantity sold. Thus, any reader who is unconvinced of the superiority of the market mass concept over actual market share can feel reassured that, for the present sample at least, the empirical differences are not large.[5]

Column 4 shows Canada/U.S. variable input productivity. One sees that there is indeed a Canada/U.S. productivity[6] gap that averages about eight percentage points. But there is no significant difference in the relative productivities of the two groups. Canadian subsectors that are relatively successful at accumulating market mass are not, on average, likely to be more or less productive in the physical transformation of inputs into outputs than are the low-mass subsectors. High productivity does not pay off in higher market mass.

Column 5 shows the difference in input prices paid (for a common bundle of inputs representing average Canada/U.S. input/output coefficients) in the two countries. van Duren et al call this the "resource cost coefficient." There is no difference in resource costs, on average, for the most and least competitive groups.

Columns 6 and 7 show profitability measures. Column 6 gives the ratio of price to cost (materials-plus-labor) in each Canadian subsector. The top-10 competitors are quite a lot less profitable than the bottom-10, though the intragroup variances are such that the difference is not significant at the 95 percent level. Column 7 shows the ratio of this to the corresponding U.S. price/cost ratio. Canada/U.S. profits do not differ much from top to bottom, and indeed are higher in the top-10 group, which indicates that Canadians tend to excel in subsectors that earn low profit margins.

Relative unit values (column 8) differ from relative prices in that the latter are index numbers, computed for a fixed basket of commodities for each subsector, whereas the unit value calculations lump together all the commodities in each country's basket and compare the average value per unit quantity, which will reflect differences in quantity shares as well as prices. If a subsector's relative unit value exceeds its relative output price, it means that the Canadian subsector has been able to concentrate more of its sales on higher-value product lines relative to the U.S. subsector.

Overall, relative unit values are higher in Canada than relative prices, but nothing much should be made of this, since the appreciation in the value of the Canadian dollar between 1986 (when the price comparisons were made) and 1988 (the year for the unit values) could have reduced U.S. prices converted into Canadian currency even if no real change in relative prices had occurred. What may be interesting is that relative unit values are quite a bit larger for the top-10 group, consistent with one of the keys to success being getting more of the output mix into higher-price commodity lines. However, intragroup variances are such that even this quite large difference in means does not show significance in a t-test.

To summarize Table 13.1: There are very substantial differences in the market mass competitiveness of Canadian food and beverage subsectors. These do not appear to be associated with some of the determinants of competitiveness that have been proposed; notably, prices, productivity, input costs, and profit margins. The search for factors that are associated with market mass is the focus of Tables 13.2, 13.3, and 13.4. To save space, only the three mean values are shown on these tables, along with an indication of the significance of the top-10/bottom-10 mean differences.

The top-10 subsectors are less likely to be U.S.-owned and much less likely to operate in advertising-intensive markets than are the subsectors at the bottom of the list (columns 9 and 10 in Table 13.2). There is little doubt that these two factors are linked. With the exception of the largely indigenous beer

TABLE 13.2 Market and Input Characteristics of Forty Canadian Industries

	9	10	11	12
	% US-Owned	High Advertising Dummy	Price Elasticity of Demand	(US) Capital Output Ratio
Mean Top-10	28	0.20	-0.35	0.36
Mean Bottom-10	44	0.90	- 0.15	0.34
Mean All	32	0.43	-0.24	0.37
Significance of Mean Differences	—	**	**	—

TABLE 13.3 Concentration and Plant Size for Forty Canadian Industries

	13	14	15	16	17	18
	Canada 4-Firm Concentration Ratio	US 4-Firm Concentration Ratio	Ratio CDA/US CR4	Size Av. Canadian Plant	Size Av. US Plant	Ratio CDA/US Plant Size
Mean Top-10	65	45	1.52	17.3	35.6	0.58
Mean Bottom-10	74	53	1.58	20.5	47.0	0.44
Mean All	69	50	1.53	19.4	41.4	0.54
Significance of Mean Differences	—	—	—	—	—	—

TABLE 13.4 Trade and Protection Measures for Forty Canadian Industries

	19	20	21	22	23	24
	Tariff on US Imports	Market Share of US Imports	Market Share of All Imports	Output Share of Exports to US	Output Share of All Exports	Non-Tariff Barrier Dummy
Mean Top-10	0.09	0.01	0.01	0.03	0.06	0.20
Mean Bottom-10	0.14	0.03	0.10	0.04	0.06	0.00
Mean All	0.11	0.02	0.07	0.04	0.06	0.18
Significance of Mean Differences	—	**	**	—	—	—

and wine subsectors, and the tobacco subsector, which has non-U.S. foreign ownership, all the highly advertised products are produced in Canada under above-average U.S. control. U.S. penetration of Canadian markets has focused on subsectors in which the U.S. parent companies have substantial investments in brand promotion. This is certainly not surprising, since advertising goodwill can be transferred at low marginal cost from the U.S. to the Canadian market, given the extensive penetration of U.S. media into the northern market and the quite similar nature of consumption technologies in the two countries. U.S. firms have a product market advantage in these subsectors which makes capital investments more attractive.

What is interesting, though, is the weak performance of the Canadian subsidiaries in building market mass compared with their parents operating in the United States. This is not simply a matter of the binational corporations choosing to serve the Canadian market with imports from south of the border, as Table 13.4 will make clear later. There is very little importing from the U.S. in these subsectors. U.S. binationals do a lot better at building market mass in their home market than their subsidiaries manage to achieve in Canada. Perhaps — as popular opinion might have it — the subsidiaries are starved of head-office resources for R&D, product promotion and so on.

So, the subsectors at which Canadians do relatively well are serving markets in which large scale, national brand-type promotional activities are not favored, and which, probably as a consequence, do not tend to be heavily penetrated by U.S. ownership. Recall from Table 13.1 that the top-10 subsectors appear to be among those that tend to operate on relatively smaller profit margins, which is also to be expected in the absence of margin-building advertising activity.

The estimates of price elasticities of demand given in column 11 are fully consistent with the advertising intensity dummies. Elasticities of the less heavily advertised top-10 group are more than twice as large, on average, as those of the bottom-10 subsectors, as would be expected in markets where price, rather than promotion, is a relatively more potent marketing tool.

These differences in product market and ownership characteristics are not reflected in differences in technology and subsector structure. The last column (12) of Table 13.2 shows absolutely no systematic differences in capital/output ratios. Concentration ratios in Canada are generally higher than in the United States (columns 13 and 14). In fact, four-firm seller concentration ratios are higher in Canada in all but two of the forty subsectors, reflecting the much smaller market size in this country. In both the United States and Canada, concentration is slightly higher in the bottom-10 group. While neither of these differences are significant, the fact that the differences occur in both countries suggests that this is a structural characteristic typical of high versus low market mass subsectors, rather than a strategic instrument generating differences in competitiveness of Canadian subsectors vis-à-vis their U.S. American competitors. A tendency towards high concentration can also be plausibly associated with the tendency towards high advertising/sales ratios in the bottom-10 group. Consistent with this, relative Canada/U.S. concentration is virtually the same in the two groups (column 15). However, these data certainly do not support the oft-mooted hypothesis that high concentration is a necessary condition for Canadian subsectors to compete in the North American market.

Canadian plant sizes are in most cases smaller than U.S. (columns 16, 17, 18), but there is some tendency for the successful Canadian subsectors to achieve higher relative plant sizes than do the bottom-10 group. In the light of all the preceding evidence, this should be interpreted as a result of superior market performance, rather than a cause of it.

Finally, Table 13.4 reports a number of measures of exposure to international competition. The results contain some surprises. First, see from column 19 that, although Canadian tariffs on imports from the U.S. are lower in the top-10 group, the difference is not significant and in fact is just about entirely due to the extraordinarily high protection offered to the least competitive subsector on the list — wine. Of the seven subsectors with prohibitive or near prohibitive non-tariff barriers to imports (column 24), two were in the top-10, five in the top-20, and none in the bottom-10 subsectors in the competitiveness ranking. That is, these numbers do not give any support to the proposition that exposure to foreign competition fosters domestic competitiveness. However, they do not prove that proposition, though the success of the much-maligned NTB subsectors is surely intriguing.

The top-10 subsectors do significantly better at deterring imports from the U.S. than do the bottom-10 group, but the numbers are all very small (columns 20, 21). In essence, there is no importing into top-10 markets in Canada, and very little, at least from the United States, into the rest of the sample. Exporting is rare throughout the sample (columns 22, 23).

What all this means is that competitiveness in Canadian food subsectors has been very much a matter of developing the Canadian market. The striking variations in position of demand curves revealed by the market mass measure far outweigh differences in import or export shares. Only two of the top-10 subsectors (malt and pork) are really substantial exporters; only two of the bottom-20 (confectionery and wineries) share a significant proportion of the Canadian market with imports.

A possible (though uninformed) response to the evident unimportance of Canada-U.S. trade in this sector is that the Canadian food economy is autarkic. Due to tariffs, transport, or other trade costs, Canadian firms are simply not subject either to pressures from import competition or opportunities to export. One can test for autarky by using the database to estimate an econometric model to explain domestic subsector pricing behavior. It turns out that, along with some other variables such as relative input prices, the Canadian tariff is a substantial and significant determinant of Canadian prices relative to U.S. Coupled with the low or zero trade shares, this result implies that Canadian subsectors tend to engage in "limit pricing"—setting prices so that imports will not be attracted into the domestic market (Eastman and Stykolt, 1967).

This means that CUSTA will have a substantial impact on Canadian prices, but one cannot say what will happen to output and competitive performance. Given the lack of an association between 1986 tariff levels and competitiveness, one might be tempted to predict that free trade will have little effect. But this is to ignore the political economy of the pre-free trade tariff structure, which probably evolved such that higher protection was afforded more vulnerable subsectors. If the North American market becomes truly integrated, elasticities facing the smaller Canadian subsectors should increase. Thus, the impact of a sizeable and across-the-board elimination of tariffs could be still be substantial. What can be said, though, is that if the market mass competitiveness index proposed here is valid as a measure of the strength of Canadian subsectors, it should predict Canada's post-free trade performance.

Finally, Table 13.5 shows the results of ranking subsectors by their input productivity (real input per unit real output) for the top and bottom groups of subsectors, when the sample is ranked from highest to lowest relative Canada/ U.S. productivity. Table 13.1 showed that there was no correlation between productivity and market mass competitiveness, so mean values for the latter variable are not shown. There is a substantial difference between the productivity performance of the two groups. The top-10 Canadian subsectors are 6 percent above par with their U.S. counterparts; the others average 30 percentage points less.

Productivity performance is not correlated with profitability. The relatively more productive subsectors show absolutely no sign of doing better in their price/cost ratios. So, if higher productivity is not pocketed in profits, it must be passed on in lower prices. This is seen in the data — relative output prices are about 17 percentage points lower in the high- productivity group.[7]

Tariffs on imports are only one-half as large, on average, for the high-productivity subsectors. This is the only variable shown on Table 13.5 (apart from productivity itself) for which the mean differences between top- and bottom-10 are significant at the 95 percent level. This could mean a causal link

TABLE 13.5 The Forty Canadian Industries Ranked by Input Productivity

Mean Values	CDA/US Input Productivity	CDA/US Price/ Cost	CDA/US Output Price	Tariff on US Imports	Market Share of US Imports	Output Share of Exports to US
Top-10	1.06	0.90	0.98	0.07	0.03	0.02
Bottom-10	0.76	0.90	1.15	0.14	0.02	0.05

from low tariffs to high productivity, as free trade proponents hope and expect, or it could indicate that low productivity necessitates high protection. Data generated from the free trade era will help resolve this issue. Note that in 1986, there was no real difference in the trade performance of high- and low-productivity groups. Low-productivity subsectors actually export more, on average, but the difference is not significant.

Conclusions

The success of Canadian food and beverage industries, relative to the benchmark of the performance of their U.S. counterparts, has been very much a domestic matter. With a few exceptions, the Canadian subsectors that have accumulated an unusual amount of market mass have done so by developing the Canadian market. Few of even the most successful subsectors export significantly, and few share much of their market with imports, though it appears that the threat of import competition does constrain price setting.

Apart from the lack of trade, the three most striking findings are (1) the lack of correlation between productivity and competitiveness; (2) the surprisingly good performance of subsectors protected by nontariff barriers; and (3) the poor showing of advertising-intensive subsectors with high degrees of U.S. ownership. Possibly connected with this is the tendency for the least competitive subsectors to be more concentrated in both countries than the average. Apart from these factors, examination of a quite comprehensive list of industrial organization variables measuring subsector characteristics failed to show systematic differences between the highest and lowest competitiveness groups. Further investigation of the very substantial range of competitiveness that has been revealed across the forty food and beverage subsectors in this sample should probably follow the case study route, focusing perhaps on a few subsectors at each end of the competitiveness ranking. It would be interesting to know, for example, why the competitiveness of the chewing gum subsector is so high and the confectionery subsector so low.

A serviceable framework for such case studies might well be the paradigm developed by Porter (1990) for at least two reasons. First, our emphasis on the importance of the subsector in its domestic market is certainly consistent with the theme developed by Porter that national or even regional strength is a key factor in fostering competitive advantage. Secondly, Porter's "diamond" of determinants of competitiveness — factor conditions, demand conditions, related and supporting industries, and structure and strategy — plus government policies offer a neat taxonomy for imposing some coherence on the mass of information that a case study can generate. A particularly important task for more intensive research is to sort out those determinants of competitiveness that are replicable (in the case of highly competitive subsectors) or fixable (for subsectors at the bottom of the ranking) from those that are intrinsic and immutable features of the market environment. Harm can be done in attempts to improve the competitive position of industries that really shouldn't show above-average performance.

The role of existing government interventions should be carefully identified and evaluated. For example, the very good performance of the Canadian butter subsector may be linked to the quite poor ranking of shortening and margarine by various government policies and programs, installed at the behest of the powerful dairy lobby, which favor butter production at the expense of its substitutes. Such policies may not be a viable or desirable model for other subsectors, and it may not be possible to benefit the margarine subsector without doing an equal amount of harm to butter producers.

This chapter presents a single measure of competitiveness, in some contrast to the eclectic, multidimensional approach espoused, for example, in the papers by Ash and Brink (1992) and by van Duren et al. (1992). The advantages of this approach are that it provides a ranking of subsector performance and offers an unambiguous target for analysis and remedies to improve competitiveness. The disadvantage, of course, is that market mass measure may be wrong or misleading as a normative measure. What this paper does is to implement empirically what is now a fairly settled definition, in Canada, of competitiveness. This paper has not offered an empirical test of the validity of the definition itself. Such a fundamental test will have to await data generated by the new, open-border, high-elasticity world of Canada/U.S. free trade.

Notes

1. Comments received from Rick Barichello, Maury Bredahl, John Connor, Larry Martin, and other participants in the consortium are gratefully acknowledged. Much of the data was collected for use in a study commissioned by Industry Science and Technology Canada. Much of the work on the present paper was carried out while the author was in the T.D. MacDonald Chair at the Bureau of Competition Policy, Ottawa. The Government of Canada cannot be held responsible for the views expressed herein. Dave Feeley, at the University of British Columbia in Vancouver, was the research assistant for the study. His work was funded, in part, by the Social Sciences and Humanities Research Council of Canada, and by the BCASCC program of the Government of British Columbia. Thanks to all.

2. This benchmark will tend to overestimate Canadian competitiveness if the overall Canadian price level is higher than that in the U.S. If so, then the benchmark should be an output ratio of 9 percent when the industry charges a price higher than the U.S. industry's price by the amount of the overall difference in relative price levels.

3. For example, see the recent paper by Connor and Peterson (1992), and the evidence gathered in the book by Connor, et al. (1985). The situation may be different for primary industries, such as farming, in which firms produce quite homogeneous products under market conditions that do not allow much market power to develop,

whereas inputs (labor, land) are heterogeneous. Thus, differences in profitability may be reliable indicators of differences in the efficiency of inputs.

4. Note that at the firm level, rather than the industry, most IO analysts would be willing to attribute superior performance at least partially to superior efficiency. That is, if a particular firm makes more money than its competitors, it is likely doing something better than its competitors. But if an industry earns persistently high profits, market power is the plausible candidate.

5. Note that this will not always hold. With larger elasticities, such as one might expect to find at the more disaggregated level of the individual firm or strategic group, differences between market mass and actual market shares will widen.

6. Variable input productivity includes labor, materials, and energy but not capital services. Capital data are not available in Canada at this level of disaggregation. Although the omission of capital services may affect the numbers for individual subsectors, it can hardly have much impact on the overall picture, since, for the food and beverage sector as a whole, capital services account for less than 10 percent of total input costs.

7. This is less than the difference in productivity means. The discrepancy can be accounted for approximately by differences in input prices — the high productivity group pays substantially more for their inputs. Whether this is fortuitous or whether higher input prices, in fact, spur superior productivity performance is an interesting question that must be left open here. Another possibility is that differences in input prices at least partially reflect differences in input quality, which in turn are reflected in productivity.

References

Ash, K., and L. Brink. "Assessing the Role of Competitiveness in Shaping Policy Choices: A Canadian Perspective." In *Competitiveness in International Food Markets*, eds. M.E. Bredahl, P.C. Abbott, and M. R Reed. Boulder CO: Westview Press, 1994.

Connor, J. M., R. T. Rogers, B. W. Marion, and W. F. Mueller. *The Food Manufacturing Industries: Structure, Strategies, Performance and Policies*. Lexington: Lexington Books, D.C. Heath and Company, 1985.

Connor, J. M. and E. B. Peterson. "Market Structure Determinants of National Brand — Private Label Price Differences of Manufactured Food Products." *Journal of Industrial Economics* 40(June 1992): 157-71.

Eastman, H. C., and S. Stykolt. *The Tariff and Competition in Canada* Toronto: Macmillan, 1967.

Harris, R. G., and D. Cox. *Trade, Industrial Policy and Canadian Manufacturing*. Toronto: Ontario Economic Council, 1984.

Kalaitzandonakes, N., B. Gehrke, and M. Bredahl. "Competitive Pressure, Productivity Growth, and Competitiveness." In *Competitiveness in International Food Markets*, eds. M.E. Bredahl, P.C. Abbott, and M. R Reed. Boulder CO: Westview Press, 1994.

Lau, L. "Technical Progress, Capital Formation, and Growth of Productivity." In *Competitiveness in International Food Markets*, eds. M.E. Bredahl, P.C. Abbott, and M. R Reed. Boulder CO: Westview Press, 1994.

Pagoulatos, E., and R. Sorenson. "What Determines the Elasticity of Industry Demand?" *International Journal of Industrial Organization* 4(1986): 237-50.

Porter. M. E. *The Competitive Advantage of Nations* New York: The Free Press 1990.

Van Duren, E., L. Martin, and R. Westgren. "A Framework for Assessing National Competitiveness and the Role of Private Strategy and Public Policy." In *Competitiveness in International Food Markets*, eds. M.E. Bredahl, P.C. Abbott, and M. R Reed. Boulder CO: Westview Press, 1994.

Wonnacott, R.J., and P. Wonnacott. 1967. *Free Trade Between the United States and Canada: The Potential Economic Effects.* Cambridge, MA: Harvard University Press, 1967.

Appendix: Variable Definitions

1. CDA/U.S. Market Mass Competitiveness Index: ratio of Canadian subsector's real output to U.S. subsector's real output if the Canadian subsector charged the same price as the U.S. subsector. (Note: all $U.S. figures are converted to $CDN at an exchange rate of 1.39.)

2. CDA/U.S. Output Price: ratio of the wholesale prices (factory gate) charged by Canadian and U.S. industries. When the subsector produces more than one commodity for which relative prices were available, these are aggregated into a basket of commodities, with shares intermediate between the shares at the commodities in Canadian and U.S. subsector output.

3. CDA/U.S. Real Output: cf. Section 3.

4. CDA/U.S. Input Productivity: cf. Section 3.

5. CDA/U.S. Resource Costs: cf. Section 3.

6. Canada Price/Cost: ratio of total value of subsector shipments to total variable (materials plus fuels plus labor) costs, Canada.

7. CDA/U.S. Price/Cost: ratio of Canada Price/Cost to U.S. Price/Cost.

8. CDA/U.S. Unit Value (1988): calculated as follows for each subsector: First collect all the commodities produced by the subsector for which both value and quantity data are available from both countries' manufacturing census. Then, for each country, add up the values and the quantities (in common units — e.g., kilograms), and divide to get unit values. Then divide the Canadian by the U.S. unit values, converted into a common currency.

9. Percent U.S.-Owned: percentage of Canadian subsector shipments from plants owned or controlled by U.S. firms.

10. High Advertising Dummy: equals one if the (U.S.) advertising/sales ratio exceeds 0.014.

11. Price Elasticity of Demand: cf. Section 3.

12. U.S. Capital Output Ratio: ratio of book value of assets to (annual) gross value of output, U.S. subsector.

13. Canada 4-Firm Concentration Ratio: percentage share of the Canadian subsector's output shipped by the 4 largest firms.

14. U.S. 4-Firm Concentration Ratio: percentage share of the U.S. subsector's output shipped by the 4 largest firms.
15. Ratio CDA/U.S. CR4: column 13/column 14.
16. Size Average Canadian Plant: total Canadian subsector value of shipments divided by number of establishments (plants) in the subsector. In millions of $CDN.
17. Size Average U.S. Plant: total U.S. subsector value of shipments divided by number of establishments (plants) in the subsector. In millions of $CDN.
18. Ratio CDA/U.S. Plant Size: column 16/column 17.
19. Tariff on U.S. Imports: ratio of total duties paid to total value of dutiable imports from the U.S.
20. Market Share of U.S. Imports: ratio of total value of (dutiable plus duty free) imports from the U.S. to value of Canadian subsector output minus exports plus all imports.
21. Market Share of All Imports: ratio of total value of (dutiable plus duty free) imports from all sources to value of Canadian subsector output minus exports plus all imports.
22. Output Share of Exports to U.S.: ratio of value of Canadian exports to the U.S. to value of Canadian subsector output.
23. Output Share of all Exports: ratio of value of Canadian exports to all countries to value of Canadian subsector output.
24. Non-Tariff Barrier Dummy: equals one if the Canadian subsector is protected by prohibitions or extreme physical limitations on competing imports.

14

Assessing the Role of Competitiveness in Shaping Policy Choices: A Canadian Perspective

Ken Ash and Lars Brink[1]

Popular and academic writers have been interested in the determinants and indicators of competitiveness for a long time. In recent years, the pursuit of improved competitiveness has attracted more attention than before, and governments have begun more explicitly to deal with competitiveness concerns in policy development. The purpose of this chapter is to explore how the notion of competitiveness can contribute to the development and reform of policies in the agri-food sector.[2] The discussion is based on recent experience in Canada.

The chapter is structured in essentially two parts. The first part reviews the Canadian government's initiative to foster improved competitiveness in the economy in general and the agri-food sector in particular and discusses the forces giving rise to this initiative. It outlines concepts of competitiveness at the sector and national levels, examines determinants of competitiveness in the agri-food sector, and investigates the linkages among these determinants and policy choices. The second part outlines the current agri-food policy agenda in Canada and the relationship between agri-food sector competitiveness and policy reform. This discussion considers sectoral structure and performance trends, the apparent objectives and impacts of current sectoral policy, and possible future agri-food policy directions.

The Policy Interest in Competitiveness

The fundamental reason for Canada's interest in competitiveness is the recognition that the world is changing, that the change is quicker than before, and that the effects of change elsewhere may now more than before be directly felt in Canada. Therefore, Canada's economy and the policies governing economic activities must also change. To resist and attempt to avoid change would be to forgo the contributions to a rising standard of living that an economy that is modern and adaptive can make. Improving competitiveness through appropriate policies is a means to an end, not an end in itself.

The end to be achieved through improved competitiveness is the continuation and assurance of economic prosperity. Key elements of economic prosperity are a domestic market place that encourages competition and innovation, a stable and attractive economic climate that encourages investment, an ability to translate new scientific and technological knowledge into high quality products at internationally competitive prices, a system of lifelong learning that provides relevant skills, and, in response to globalization trends and to the integration of domestic and international policies, improved access to foreign markets (see, for example, Government of Canada, 1991).

Over the last few years, the Canadian government has explicitly brought the pursuit of improved competitiveness into its discussion of the appropriate orientation of public policy. In 1991, an extensive series of consultations at the national, sectoral, and community levels were launched with the announcement of the Prosperity Initiative. The broad objective of this Initiative is to explore how government and industry could work better together to ensure improved industry competitiveness and economic prosperity. Earlier work, which examined the nature and role of competitiveness in public policy, contributed to the focusing and articulation of issues in the "Prosperity Through Competitiveness" and other discussion papers that were prepared to help guide these discussions. Some examples of writings that have been instrumental in giving currency to competitiveness issues as a public policy concern in Canada are D'Cruz and Rugman (1991), Economic Council of Canada (1992), Industry, Science and Technology Canada (1991), Porter (1991), Purchase (1991), and Science Council of Canada (1992). The Canadian government also participated with a major private sector group in funding an examination of Canada using the approach developed by Porter (1990).

An essential characteristic of improved competitiveness as a policy objective is that there is no one single policy instrument that will bring about improved competitiveness. Rather, the role of government policy is to ensure that the social and economic environment in which firms and sectors operate is conducive (or at least not detrimental) to improved competitiveness. This means that an improved understanding of competitiveness as a policy objective and of the determinants of competitiveness permeates the considerations of all policy — the reconsideration of existing policy frameworks as well as the development of new policy initiatives. It also means that policy makers increasingly see government as a facilitator and stagesetter (as opposed to creator), they increasingly view competitiveness as something to be achieved over the long-term (as opposed to "buying" market share), and they are increasingly receptive to considering economic sectors in the traditional sense as parts of a more complex set of economic interdependencies, such as links in a whole chain of value-adding activities.

A major force that has brought the pursuit of improved competitiveness onto policymakers' agenda in Canada is concern about the slowdown of growth in real income since the mid-1970s due to slow productivity growth. The decline in Canadian productivity growth is, in turn, associated with problems in such areas as developing and using new technologies, providing and taking advantage of training and education opportunities, making domestic markets work in ways that promote growth and employment, linking suppliers and users of investment capital, and building partnerships among governments, labor, and business (Government of Canada).

In the area of international trade, specifically, there is concern about how Canada fares in the implementation of the trade agreement with the United States and about lacklustre performance in exports (falling market shares in many export markets, especially outside the United States). Although not so much an issue in trade internationally, the existence of barriers to trade between provinces within the country is also a concern in Canada.

Other forces are also pushing the notion of competitiveness to the fore — the same forces that are making themselves felt in other countries. Important forces in this regard are the new exchange rate regimes adopted by many countries in recent decades and the ease with which capital now moves among countries. They also include the accelerating pace of change brought on by improved communications and rapid scientific advancement, the creation of a more global economy, with transnational corporations increasingly shaping trade flows across borders through their production/location decisions, and the growing importance of knowledge, rather than raw materials, to the creation of wealth. Moreover, more attention is now paid to such indicators of economic performance as the extent of the upgrading of skills and resources, in addition to traditional ones, such as the size, direction, and rate of change of the balance of trade.

In agriculture or, more appropriately, in the chain of value-adding activities in which farming is one link, there are more specific forces of change that require policymakers to reconsider both their objectives and their tools. The characteristics of market demand are changing (for example, increased demand for products with more value-added and falling demand for butterfat are important developments in the dairy subsector). New methods and techniques using new inputs are being adopted both in farming and in food processing — sometimes yielding productivity increases that are large in relation to demand increases but still low in relation to productivity increases achieved in other sectors or in other countries. The climate of fiscal restraint contributes to a renewed need to reconsider how much public funds are spent and how they are spent: Public funds might be better used in facilitating adjustment to change than in helping to resist change.

Agricultural trading patterns will change as a result of differences among countries with regard to rates of population, income and productivity growth; differences between commodities in income elasticity of demand; and differences between rich and poor countries in income elasticity of demand of a given commodity (McClatchy and Warley, 1991). Fundamental changes in the functioning of the economy in a number of countries (such as countries in Eastern Europe and the former Soviet Union) affect production, consumption, and trade. Reform of agricultural and trade policies resulting from a conclusion of the Uruguay round of trade negotiations will further change the nature of the agri-food sector in all parts of the world. Trade liberalization and agricultural policy reform may also continue independently of the Uruguay Round, perhaps more as a result of change in the global economy than as the fundamental cause of change.

In order better to bring an appreciation of the consequences of change in the world to bear on the development of agri-food policy in Canada, the Canadian government launched an Agri-Food Policy Review in 1989. One of the government-industry task forces established under the review was the one on Competitiveness in the Agri-Food Industry. It was charged with assessing impediments to improved domestic and international competitiveness in the Canadian agri-food sector. While the recommendations of the task force focused mainly on changes that should be initiated by the agri-food industry itself, the role of government in nurturing competitiveness was acknowledged. The work of the Task force led to the establishment of the Agri-Food Competitiveness Council, an all-industry group whose mandate is to advise federal and provincial agriculture ministers on ways to improve the competitiveness of the sector.

The deliberations of the Task Force on Competitiveness in the Agri-Food Industry centered on the four pillars established for the Agri-Food Policy Review: (1) more market responsiveness, (2) greater self-reliance in the agri-food sector, (3) a national policy that recognizes regional diversity, and (4) increased environmental sustainability. The task force also considered basic changes in the role of government from that which had become expected or accepted over the last few decades. The pillars amount to a set of de facto policy objectives, although they are more conceptual than operational. As policy objectives, they provide some of the dimensions in which policy initiatives should be assessed. Government interest in competitiveness, in many ways, represents an integration of these policy objectives into a comprehensive statement of the policy direction to be followed. Such a policy direction provides a common aim for the further evolution of policies specific to or related to the sector.

The Concept of Competitiveness

If competitiveness is to be meaningfully addressed by policy initiatives, all participants in the policy development process must have a reasonably precise and common understanding of the concept at the sectoral and national level. The following two definitions are relevant.

The first definition (from a working definition developed by the Task Force on Competitiveness in the Agri-food Industry) states that a competitive industry is one that has the sustained ability to profitably gain and maintain market share in domestic or export markets. This definition highlights not only profitability but also, through its emphasis on market share, the relation between the size of the market and the size of the industry measured in terms of output. Policy priority is often linked to changes in output of an industry because many factors affecting output also affect value-added or the industry's contribution to the economy. However, a definition emphasizing market share must be used carefully. For example, in a situation where an industry's output is increasing, but total markets are increasing more rapidly, the industry's market share declines. Applying the market share definition to this industry would thus identify it as not competitive.

The second definition is derived from a number of definitions proposed over the last few years: An internationally competitive industry or sector is one that has the sustained ability to deliver goods and services in the form and at the time, place, and price that together form a package at least as attractive to buyers in domestic and international markets as that offered by potential foreign suppliers, while earning at least opportunity costs on resources employed.

This statement of what competitiveness means puts the emphasis on the package of price and other attributes of the product. It emphasizes the conditions that must be satisfied in order to be competitive, especially the need to meet buyer preferences for various attributes of the product and the relationship of cost to other uses of the inputs, while retaining the elements of sustainability and profitability. This definition thus helps to identify determinants of competitiveness. It also avoids some of the issues in using a market share definition, such as the one arising when markets grow faster than output. However, other problems remain, for example, the issue of how government programs and barriers to trade affect indicators of competitive performance.

For industry, competitiveness implies the ability to adapt, innovate, identify, and exploit business opportunities. This dynamic sense of the concept is clear from the common elements in a number of definitions developed for analysis at the sectoral or national level. It becomes particularly clear when the definitions are paired with the explicit recognition that changes are taking place ever more rapidly in the national and global economy of which the agri-food sector is a part.

For governments, facilitating the improvement of competitiveness in a sector requires putting in place policies that do not overly stifle individual initiative but rather enable or even encourage individuals and firms to deploy resources in response to economic, market-driven incentives. Fostering competitiveness is a means to achieve a higher standard of living. As such, it requires a long-term planning horizon and recognition of the essential role played by structural adjustment, in the economy and within individual economic sectors, in contributing to economic growth (see, for example, Organization for Economic Cooperation and Development, 1987).

Determinants of Competitiveness and Policy Choices

The Porter exposition and analysis of competitiveness in relation to government policy and the work of others, such as D'Cruz and Rugman (1992) and the IMD and the World Economic Forum (1990), attempt to address points that traditional economic theory (such as partial equilibrium analysis) approaches only with difficulty, if at all. For example, it draws attention to the roles played by upgrading of factors, development of new products, and clusters of industries. This is the result of insights from traditional and new trade economics, strategic business management, and industrial organization. Theories focusing on transaction costs and public choice have potential to further clarify important linkages, such as those among sector structure, government policy, and competitiveness.

In highlighting the determinants of competitiveness, some recent work (e.g., Porter [1991] and D'Cruz and Rugman) moves beyond analysis of size of market shares and how market shares change over time to focus explicitly on why change has occurred. It sheds more light on the whole set of factors that determine competitiveness and on the explanations for particular competitive performance than can be derived from analysis of single factors alone, such as cost comparisons or analysis of changes in trade policy.

Analysis of various measures of domestic and export market shares, performance measures (e.g., rates of growth of output, value-added, investment) and other indicators of competitive performance is of course useful in getting a perspective on where an industry is doing well and not so well in the first place (such as the work of Haley and Krissoff [1986], Paarlberg et al. [1985] Vollrath [1987, 1989] and Vollrath and Vo [1988]). Such analysis shows promise in linking competitive performance with particular policy directions, but overall it appears difficult to come up with clear policy priorities on the basis of analyzing competitive performance indicators alone. A major hurdle in doing so is the need to take account for trade barriers and subsidies at home and abroad. For example, market share data for the highly protected Canadian dairy and poultry subsectors provide no information on their competitive performance or any indication of policy changes needed.

For policy purposes, the issue is not so much how competitive an industry is but how it might be made more competitive. Assessments of competitive performance can be helpful in identifying those parts of the sector that may require particular attention in policy development. Analysis of the determinants of competitiveness provides the understanding that is crucial to the development of policy to improve competitiveness.

Pursuing Improved Competitiveness

If improved competitiveness was the only policy objective to pursue, making policy choices in the agri-food sector would be relatively simple. However, macroeconomic, social, regional development, and competition policy objectives, for example, provide further direction when making choices in agri-food policy. Moreover, because of the connections and interdependencies among economic sectors, any particular policy objective (including that of improved competitiveness) set for another sector will have consequences for the policy choices in the agri-food sector.

Given that several policy objectives can be in place to guide the policy choices for the agri-food sector, all such choices will not be on a convergent course towards increased competitiveness at all times. Changing policy objectives and different weights given to particular objectives at different times make the convergence of policy choices an illusory expectation. Progress in a generally desired direction may follow a zig-zagging path: policy choices at times being pulled more towards one objective and at other times pulled more towards another objective. Setting a common direction for many sectors and policy areas, such as the direction of improved competitiveness under Canada's prosperity initiative, facilitates making policy choices such that they do converge on a common path.

Pursuing improved competitiveness requires a focus on the agri-food sector as a whole and on the relationships between its component parts. Concentrating on just one part of the sector, such as production of commodity X in region Y, may improve the performance indicators of that part but only at the expense of some other part of the sector. This approach tends to foster antagonism between subsectors and to impede change. A more fruitful approach might be to reduce unnecessary friction between subsectors, to avoid creating new friction and to firmly keep a focus on the competitiveness of the agri-food sector as a whole, while recognizing that it is only one part of the overall economy.

A Discussion of the Agri-food Policy Agenda in Canada

The following discussion provides a snapshot of the agri-food policy agenda in Canada by highlighting recent structural and performance trends, current policy objectives and impacts, and possible future policy directions. An attempt is made to illustrate the contribution of competitiveness in shaping policy choices.

Evolution of the Agri-food Sector: Structural and Performance Trends

The agri-food sector in Canada, as in many other countries, has shown its capacity to respond to the stimuli generated by the combination of markets and public policy. The agri-food sector in Canada has been, and continues to be, an important part of the overall economy. The farm production and food processing segments of the agri-food sector contribute about $24 billion or almost 5 percent of Canada's total gross domestic product. Agriculture is the second largest primary sector behind the mining, quarrying, and oil well sector, while food and beverage processing is the second largest manufacturing sector behind the transportation equipment sector.

The nature of the agri-food sector continues to change. Productivity-enhancing technological advances serve over time to reduce commodity prices. They also increase the set of resources that one person can manage, which, in the absence of new markets opening up, implies a need for ongoing sector adaptation, including the out-migration of labor and businesses. The alternative to such adaptation would, in most circumstances, be an income problem (it would not be possible to maintain the business revenue levels required to adequately remunerate all resources). With fewer and larger farms and a much-reduced farm population, agriculture could not sustain the viability of many smaller, rural communities. Small, regionally-oriented processing plants will not easily achieve the efficiency needed to respond to market demand in terms of price, quantity, quality, and variety. While there is a wide range in firm size, Canadian firms tend to be much smaller than their U.S. counterparts.

This is not to suggest that there is no place for small or specialized farms and processors. But in light of increasing international competition, smaller operations will have to be extremely innovative if they are to remain competitive and earn sufficient income from the market. Such innovation may take many forms, from specializing to exploit niche market opportunities to diversifying product mix or income sources so as to exploit regional, seasonal, or other opportunities.

As a country that is small in terms of population but rich in natural resources, Canada's economic well-being has been closely tied to exports of its resource industries and resource-based manufacturing industries. For example, in 1989, exports from resource-related industries accounted for about one-half of Canada's total exports, compared with about one-third in both the United States and the European Community (Economic Council of Canada, 1992).

The agri-food sector itself is also heavily oriented towards exports. Exports of processed food and beverage products represent about 10 percent of total production and the corresponding share for primary agriculture products is almost 50 percent. It is, therefore, simply not feasible to close Canada's borders to all agri-food imports and have the agri-food sector serve only the domestic market (and this is apart from any other trade policy considerations such a move would entail). A larger market beyond the domestic one is necessary to sustain a Canadian agri-food sector of the current size or even of a smaller size.

The importance of export markets for the agri-food sector will further increase as new productivity-enhancing technology continues to reduce the quantity of inputs needed to produce the same quantity of output. This is significant in terms of the labor resource, but there are also other advances that can have further implications for industry structure. For example, use of growth stimulants for hogs yields more saleable meat per hog. With fewer hogs needed to meet unchanged or even strengthened demand for pork, the number of hog producers might also decline, *ceteris paribus*.

Relatively stagnant domestic food consumption in Canada (due to the relatively high income and modest rate of population growth) will require fewer and fewer producers and processors as the output of each firm increases. On the other hand, there are substantial opportunities for exports to regions, such as parts of Asia, where population and income growth combine to create increasing demand.

Current Agri-Food Policies and Programs

Direct government expenditures by departments of agriculture in Canada on food and beverage processing are relatively minor compared with such expenditures directed to farms. However, general services provided by governments to processors, such as food inspection and research and development, can be significant. Various other departments and agencies provide financial assistance to food processors in the form of grants, contributions, loans, and tax incentives. The magnitude of these expenditures relative to the size of the industry is not clear. Various measures of government intervention at the farm level indicate clearly that current public efforts are focused primarily at supporting farm incomes and that support levels vary substantially across commodities and sizes of farms. Analysis of public accounts data shows that approximately 66 percent of federal spending on agriculture in Canada can be described as directly supporting farm incomes, 21 percent is related to research and inspection services, 6 percent to development and adjustment assistance, and 8 percent to administration.

A more complete perspective than that given by data on government expenditures alone is given by data underlying more inclusive measures of support, such as producer subsidy equivalents (PSE). PSE accounts for both government expenditures and the transfers to producers generated by regulations (such as border measures) that raise domestic prices above those prevailing in international trade. Figure 14.1 shows the evolution of major components of PSE for Canada from 1979 to 1991.

There are two striking features to the evolution of PSE components (Figure 14.1). One is the variability from year to year in government expenditures on what is called direct payments (for Canada, this category includes such programs as stabilization programs and crop insurance) and ad hoc income support programs. The other striking feature is the distribution of support among types of programs: Various kinds of income transfers, such as market price support and direct payments, account for the overwhelming majority of transfers to producers (about 90 percent), while transfers in the form of expenditures on general services account for very little of the total transfer (about 10 percent).

"General services" consist of expenditures on such items as research, training, extension, inspection, and development of infrastructure in agriculture — in other words, expenditures expected to contribute to improved competitiveness of the agri-food sector by improving its capability to innovate, adapt, and exploit emerging new opportunities. That so little is spent on such programs, compared to the large amounts spent on supporting incomes directly, is a concern in agri-food policy discussions in Canada.

In summary, most of the support to producers of supply-managed commodities (milk, poultry meat, eggs) is by way of regulations, which maintain prices above world prices. Support is therefore proportional to output. In the case of grain and oilseeds producers, beneficiaries of most government support are large, higher income producers; smaller, lower income producers benefit less relative to their proportion of the farm population but more relative to their proportion of output. Little incentive is provided to encourage producers to adapt to changing market and economic conditions, compared to the support designed to maintain incomes in the short term.

Agri-food Policy Reforms: Possible Strategic Directions

The structure and performance of the agri-food sector in Canada today reflect policies, institutions, and attitudes that have evolved over many years. A desire to serve the domestic market, along with goals of self-sufficiency in production and processing in particular provinces or regions (sometimes related to maintaining or creating employment), has characterized much of the public involvement in dairy, poultry meat, egg, horticulture, and, to a lesser extent, red meat subsectors. The grain and oilseed subsector shows, in large part, the effects of policies that see the prairies in Western Canada as a major world supplier of unprocessed agricultural commodities. Continuing structural and performance problems in the sector, despite industry adjustments that have taken place, are evidenced by the considerable and increasing reliance of the sector on public income and price supports. Moving away from this situation will require changes to many traditional policies, institutions, and attitudes in Canada.

A key objective of future agri-food policy might be to facilitate the efforts of the sector to adapt to changing economic and market circumstances so as to

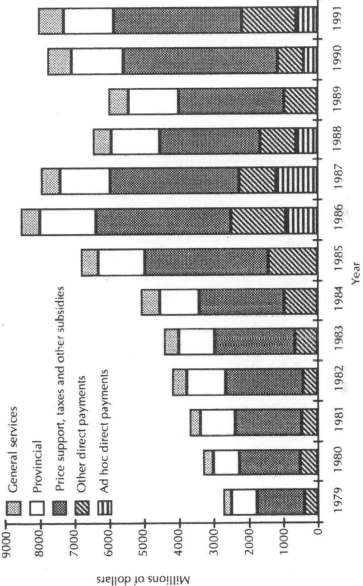

Source: Developed from OECD Producer Subsidy Equivalent (PSE) data, March 25, 1992, Farm Support and Adjustment Measures II, Canadian Crop Drought Assistance Program.

FIGURE 14.1 Structure of Support to Canadian Agriculture, 1979-91

improve international competitiveness. Policies to achieve this objective would be characterized by more market responsiveness, greater self-reliance in the agri-food sector, recognition of regional diversity, and increased environmental sustainability. To achieve this objective, a two-track approach might be appropriate:

1. Gradual reform of major existing sectoral policies, including a gradual shift in type of government expenditure away from income support and towards initiatives that support adaptation and improved competitiveness, and

2. Introduction of complementary measures that assist industry efforts to improve profitability and rely less on policy support.

Policy reforms would thus focus on removing impediments to adaptation and improved competitiveness, including the impediments that have been created by governments. The following outlines broad policy areas and possible future directions that might be pursued in these areas. It must be emphasized that any changes are directional and are considered here for discussion purposes only.

Farm Income Support and Safety Nets. Government expenditures to support farm incomes are channeled through a variety of measures — some of these are financed jointly by federal and provincial governments and farmers, including the Gross Revenue Insurance Program (Revenue Protection Component and Crop Insurance), the National Income Stabilization Account and the National Tripartite Stabilization Program. In recent years, ad hoc income support payments have also been used extensively.

The long-term aim might be a national, resource-neutral, and commodity-neutral safety net that stabilizes income (without permanently replacing part of market income) and that is equitably cost-shared among governments and farmers. A first step towards a revision of safety net programs and supply management schemes could be the assignment of a small portion of tripartite safety net premiums and supply management levies to create an industry-controlled fund to support adaptation and competitiveness-enhancing initiatives. Such a change would parallel recent changes to Canada's Unemployment Insurance (UI) system and the reallocation of some UI monies from passive income support to active training and job development measures. The approach would send a clear signal to industry, placing both the problem and the solution more in its hands.

Supply Management. Current policies in Canada limit domestic production or marketing through quotas, control imports, regulate the setting of prices, and provide more stable returns for producers and primary processors of five commodity groups (milk, chickens, turkeys, eggs, and broiler hatching eggs). This group represents about one-quarter of farm cash receipts.

The long-term aim might be a system that better responds to changes in market demand as they are conveyed through the chain of consumers, distributors, and processors. This might be achieved through institutional changes to provide for more equitable representation of all links in the chain and to increase policy transparency. Introducing more flexible pricing procedures for products used in further processing and more opportunity for transferring production (or marketing) quota across regions in response to changing market demand might also be appropriate steps towards increasingly responsive supply management schemes. New measures to encourage sector adaptation and improved competitiveness could be funded by supply management levies.

Compensation and Adjustment Policy. In the absence of an explicit adjustment policy, government has responded to various commodity pressures (for example, from producers of tobacco, grapes, and wine) with initiatives to encourage adjustment out of the production of these commodities. Some such initiatives have involved government expenditures that appear large in relation to the size of the target group.

The long-term aim of these initiatives might be a clear and transparent policy that, in the case of structural or long-term performance changes in the agriculture sector, provides appropriate support to aid in the transition to new farm or non-farm means of earning income. Measures tailored to the needs of specific commodities or industry segments might be developed while keeping in mind such guidelines as maintaining a perspective on the agri-food sector as a whole and not introducing new, unnecessary distortions.

Grain Transportation Policy. Current policies compensate the railways for a portion of the cost of shipping eligible crops from Western Canada's prairie region to eastern Canada and other export points. The effect is an upwards shift of the supply curve faced by users of these commodities in Western Canada, with consequent less value-adding (livestock feeding, processing) in that region than otherwise would be the case.

The aim might be to reduce disincentives in using Western Canada's grain, while maintaining the benefit of current spending levels for the grain economy in that region and improving transportation system efficiency. This might be achieved through reform packages with broad acceptance across industry and governments. A number of options are being actively pursued by governments.

Regulation. A review of Agriculture Canada regulations, encompassing food standards and safety, and the operation of the Canadian Wheat Board and the Canadian Grain Commission was initiated in the spring of 1992. The long-term aim is less and better (more efficient and more effective) regulation, including greater industry involvement in monitoring and enforcement. This could involve removing regulations that are obsolete or that clearly and unnecessarily constrain industry's ability to compete effectively and developing an effective framework for screening possible future regulations. The application of a "competitiveness test" as a component of the screening process is part of the review of regulations.

Research, Development, and Technology Transfer

Public research and technology transfer efforts in Canada focus on environmental sustainability, food safety, food and non-food product diversification, improved production practices and methods, and application of new technologies. Industry funds only about 16 percent of research and development in the agri-food area. The aim is to ensure rapid and effective technology transfer, to ensure research priorities are more market driven, and to build the interest and ability of the private sector to increasingly carry out its own research. Quicker and more widespread industry use of available technologies, developed at home or abroad, might be facilitated without reducing the incentives for industry to generate technological advances. Enabling producers collectively to generate funds to be used for new research and development activities might be another priority.

Trade Policy. Trade policy strives to secure improved access to global markets and a strengthened framework of equally-applied international trade rules. In the face of accelerating change in the global environment of Canada's agri-food sector, the importance of continuing this pursuit is clear.

Complementary Measures. A number of measures could be designed to complement the above examples of policy reforms and further enhance the sector's ability to adapt and compete more effectively. Funding for these measures could be provided from monies set aside from established safety net programs and supply management schemes. Such a reorientation of spending is clearly more desirable than only an increase in spending.

Setting monies aside for initiatives to foster adaptation and improved competitiveness would facilitate the development of two types of measures. First, national level programs could provide the essential support services that are needed across the country, and across the sector, if components of the agri-food sector are to be better equipped to adapt and compete. Second, regional level programs could provide more specific support targeted to commodity or geographic areas where major adjustments are needed or where major development opportunities arise. In practice, of course, national and regional programs tend to both complement and overlap each other.

The following outlines some possible initiatives at the national level to provide essential support services. Additional measures might be considered in light of circumstances prevailing for specific commodities, industry segments, regions, or provinces.

A renewed effort in the area of human resource development might improve skills, knowledge, and abilities among managers and workers. At the farm level, the variability in returns among otherwise similar operations gives evidence of the need and potential for improved management. At the processing level, slow technology adoption and low rates of productivity growth demonstrate a need for development of knowledge and skills between management and labor. Perhaps of most interest would be the provision of assistance to low-income families for business development and for training and seeking employment. Such assistance would enable them to improve income from non-farm sources. Off-farm income is the only realistic option for many families on small farms, whether they remain farming part-time or eventually give up farming.

Initiatives to improve business linkages and business arrangements might improve the economic performance of the sector. The efficiency and effectiveness of vertical linkages are a function of how well information, services, and products are transferred up and down the system. As market demands change, and the technological capacity of parts of the value chain to respond also changes, vertical coordination systems may need to adapt as well. Assessing current and possible new business arrangements and institutions might be a priority, leading to policy proposals that would facilitate the emergence of new and more adequate arrangements and institutions.

Canada's export performance is declining in many regions outside the United States and particularly in markets for higher-valued agri-food products. Steps might be taken to facilitate industry efforts to serve appropriate export markets. A trade development initiative might be designed to strengthen industry trade associations, obtain and better use foreign market intelligence, improve exporter-customer relations, and generally increase the trade readiness of the private sector.

The risk of developing and testing new products, new uses for products, and new processes at the farm level (e.g., new varieties and new cropping and breeding systems) and at the processing level (e.g., new packaging, new further processed products, and new non-food uses for traditional commodities) could be reduced through some type of innovation program. The objectives could be to diversify farm-level production and to stimulate value-added activities in response to an identified market demand. Assistance could take the form of technical advice or venture capital.

Conclusion

Improving international competitiveness is an important policy issue in Canada. Although income levels remain high by world standards, income growth has declined, trade performance has been weak, productivity growth has been relatively low, and various economic and business reports have highlighted perceived weaknesses in Canada's resource industries, research and development performance, education systems, etc. These concerns are heightened by global trade liberalization and, in particular, by the implications of Canada's free trade agreements with the United States and Mexico.

The current policy interest in competitiveness of the agri-food sector is a forward-looking one. To address questions on competitiveness, research and analysis are needed to better understand:

- how the sector is performing and might perform in the future as technology changes, markets shift, trade barriers erode, etc. Indicators of performance (and competitive potential), however measured, can be useful to suggest areas where the sector has a development problem or opportunity.
- the determinants of industry competitiveness — why the sector performs (or might perform in the future) well or poorly, better or worse, relative to competitors and relative to its own current or previous performance.
- the impacts of existing sectoral and nonsectoral policies, and alternative new policies, on the determinants of competitiveness and on the ability of the sector to adjust and respond effectively to changing market and economic signals.

Research and analysis must move quickly beyond measuring indicators of competitiveness towards explaining why some subsectors become more or less competitive, how a competitive strength was or can be achieved, and how a competitive weakness could be dealt with. Agri-food policy must increasingly aim to reorient efforts away from passive farm income support and towards enabling industry to adapt and exploit its strengths relative to emerging economic opportunities.

Improving competitiveness brings a new and perhaps more applied perspective to the debate on agri-food policy change. It helps to set direction, by emphasizing the importance of industry self-reliance and of all parts of the sector working better together to respond, profitably, to market demands; it embraces change and innovation, built on sectoral strengths and economic opportunities; and it recognizes the role of government in establishing the broad operating environment for business, but places the onus on industry itself to act.

Development of the social and economic environment in which Canada's agri-food sector can best realize its competitive capability, internationally, will require substantive shifts in attitudes and approaches. This will not happen overnight, but we can begin by developing a mix of public policies and programs that provide incentives to the sector to adapt to changing market and economic conditions and also improve the sector's capability to do so. The ultimate aim is not increased competitiveness for its own sake; the aim is sustainable income and employment growth and improved living standards for Canadians.

Notes

1. Economists in the Competitiveness Division, Agri-Food Policy Directorate, Agriculture Canada, Ottawa K1A 0C5. The views expressed are those of the authors and do not necessarily reflect those of the Government of Canada.

2. "Agri-food" is used here as a contraction of agriculture and food, i.e., the sector is taken to encompass both farming and food processing, with associated upstream and downstream activities. It is not synonymous with just that part of the food processing industry that gets its raw material from agriculture.

References

D'Cruz, J. R. and A. M. Rugman. "New Compacts for Canadian Competitiveness." Commissioned by Kodak Canada Inc., Toronto, Ontario, March 1992.

Economic Council of Canada. *Pulling Together: Productivity, Innovation and Trade*, Cat. No. EC22-180/1992E, Minister of Supply and Services Canada, 1992.

———. *Prosperity Through Competitiveness.* Consultation Paper, Cat. No. C2-177/1991E, Minister of Supply and Services Canada, 1991.

Haley, S. L., and B. Krissoff. "The Value of the Dollar and Competitiveness of U.S. Wheat Exports." U.S.D.A. Economic Research Service Staff Report No. AGES860611, July 1986.

Industry, Science and Technology Canada. *Industrial Competitiveness—A Sectoral Perspective.* Consultation Paper, Cat. No. C2-191/1991E, Minister of Supply and Services Canada, 1991.

IMD and the World Economic Forum. *The World Competitiveness Report, 1990,* Lausanne/Geneva, 1990.

McClatchy, D., and T. K. Warley. "Agricultural and Trade Policy Reform: Implications for Agricultural Trade." Paper presented at the XXI Congress of the International Association of Agricultural Economists, Tokyo, August 22-29, 1991.

Organization for Economic Cooperation and Development. *Structural Adjustment and Economic Performance*, Paris: OECD, 1987.

Paarlberg, P. L., A. J. Webb, J. C. Dunmore, and J. L. Deaton. "The U.S. Competitive Position in World Commodity Trade." In Agricultural-Food Policy Review: Commodity Program Perspectives, USDA Agricultural Economic Report, No. 530, July 1985.

Porter, M. E. *The Competitive Advantage of Nations*, New York: The Free Press, 1990.

———. "Canada at the Crossroads, the Reality of a New Competitive Environment." Business Council on National Issues and Minister of Supply and Services, October 199`.

Purchase, B. B. "The Innovative Society: Competitiveness in the 1990s." *Policy Review and Outlook* Toronto: C.D. Howe Institute, 1991.

Science Council of Canada. *Reaching for Tomorrow*. Science and Technology Policy in Canada 1991, Cat. No. SS31-22/1992E, Minister of Supply and Services, 1992.

Task Force on Competitiveness in the Agri-Food Industry. "Growing Together," Report to Ministers of Agriculture, Ottawa, June 1990.

Vollrath, T. L. *Revealed Competitive Advantage for Wheat*. USDA Economic Research Service Staff Report, No. AGES861030, February 1987.

———. "Competitiveness and Protection in World Agriculture." Agriculture Information Bulletin, No. 567, July 1989.

Vollrath, T. L., and D. H. Vo. "Investigating the Nature of World Agricultural Competitiveness." USDA Economic Research Service Technical Bulletin, No. 1754, December 1988.

15

Assessing the International Competitiveness of the New Zealand Food Sector

Ralph G. Lattimore[1]

In 1989, the New Zealand Trade Development Board commissioned a study on competitiveness of the economy by the Harvard Porter Project Team. Its report was published in 1991 and entitled *Upgrading New Zealand's Competitive Advantage* (Crocombe, Enright and Porter, 1991). This so-called Porter Report (PR) is interesting from a number of perspectives. The report is very normative by design and its conclusions rest strongly on the following statement: "The primary source of competitive advantage for the majority of New Zealand export industries that make up the vast bulk of our exports is our favourable natural- factor conditions complemented by efficient production. These industries compete mostly on the basis of low-cost primary production that relies on basic factor advantages" (Crocombe et al., p. 95). The report is referring to the natural factors of climate, soils, fish stocks, resources, water, and topography which advantage agriculture, forestry, fishing, and hydroelectricity generation (and aluminum smelting).

With this as a basis, the report goes on to appraise the efficiency of the resource generation, production, marketing, and exporting system. The authors are very critical of the single-desk selling arrangements that are frequently used in the New Zealand agricultural sector (dairy products, kiwifruit, apples, pears, other horticultural products, beef, and mutton). Other presumed policy or institutional deficiencies are identified by the authors in regulations preventing labor market adjustment and the lack of coordinated development in human resources, technology, and other areas.

The PR concludes that "The New Zealand economy is not well suited to the imperatives of the modern global economy. Despite recent reforms, our economy has continued to languish. The weak competitive position of many of our industries remains essentially unchanged. Government spending has become an even larger drain on the national economy." The report has a number of other features. In contrast to its parent work, "The Competitiveness of Nations" (Porter), PR devotes relatively more attention to the influence of government policy than it does to corporate strategy — it is much more an economic policy document than a study in industrial organization. For example, in the final chapter, PR has seven pages of recommendations for firms and fourteen pages of recommendations for the government. Nevertheless, PR is a starting point for a paper on New Zealand's international competitiveness in agriculture. New Zealand is a very large agricultural exporter relative to its geographic and population size. Like many other countries, New Zealand has its own mythology regarding agricultural trade competitiveness. The objective here is to clarify the sources of the real strengths in New Zealand agriculture and to identify the weaknesses.

The Economy and Place of Agriculture

The geography, history, and culture of New Zealand have a number of unique features that affect agriculture. This island nation is approximately the size of Japan or the United Kingdom (270,000 km²) and is situated between the 40th and 50th parallels in the South Pacific, 2000 km from its nearest neighbor, Australia. The climate is temperate, the rainfall moderate, and the topography very uneven. Until 1973, New Zealand was very closely tied commercially and politically to the United Kingdom, its major market half the globe away. Its physical isolation imposes high costs on trade and great advantages in terms of pest and disease barriers.

The pragmatic traits of New Zealand culture and institutional development have produced valuable targeted technological developments in diverse agricultural enterprises, including dairy and pastoral farming, apples, kiwifruit and other fruit and vegetables, deer, goats, and racehorses. The particular climate and topography have stimulated unique farming systems. Production systems are also often less intensive in environment-damaging inputs, which offers opportunities in the current market environment.

The economic linkages between the agricultural industry and the rest of the economy are relatively strong. Farm-based products accounted for 57 percent of total merchandise exports in 1991 (having fallen from 80 percent in the 1960s). The agricultural sector is somewhat larger than in other industrialised countries, representing around 6 percent of gross domestic product (GDP), but with agricultural input supply and agricultural processing industries contributing another 6 percent. Agriculture contributes more than proportionately to employment in the economy at the processing level. The short- and long-term financing requirements of the agricultural sector and its allied industries are large relative to the size of the capital market, both because of its capital intensity and its GDP contribution (Rayner and Lattimore, 1991).

The principal components of New Zealand agricultural production are given in Table 15.1. The largest subsectors include dairy, cattle production, and sheep production; but increasingly fruit (including kiwifruit and apples) and vegetables are prominent in the total. By contrast, crop production is relatively small, reflecting topography as well as technological advances in other areas. Agriculture has been able to maintain a place at the leading edge of development, which arguably exceeds its size, either in terms of production or trade share.

New Zealand appears to have a strong comparative advantage in selected agricultural products — especially livestock and horticultural products. Agricultural exports per capita are high by world standards (Table 15.2), particularly prominent are processed products, meat, dairy products, and wool. The relative size of these exports in relation to the total is shown in Table 15.3. Unprocessed agricultural products tend to be exportables, including fruit and vegetables and specialty seed products. A number of agricultural products tend to be importables in New Zealand, including wheat, rice, oilseeds, pigmeat, and selected fruits and vegetables.

TABLE 15.1 Gross Agricultural Production in New Zealand (million $US)

	Year Ended 31 March 1990
Wool[1]	1,235
Sheep and Lambs[2]	865
Cattle[2]	1,240
Dairy Products	2,166
Pigs[2]	125
Poultry and Eggs	215
Crops and Seeds	338
Fruit and Nuts	623
Vegetables	412
Other Horticulture	166
Other Farming	259
Agricultural Services	620
Value of Livestock Change	131
Sales of Live Animals	721
Total Output	9,117
Agriculture as a Percentage of Gross Domestic Product	6.2

[1] Excludes slipe wool and sheepskins.
[2] Sales for slaughter, including on-farm kill.

Source: Ministry of Agriculture and Fisheries (1992).

TABLE 15.2 Per Capita Agricultural Trade, Selected Countries 1989 ($U.S.)

	Agricultural Exports Per Capita	Agricultural Imports Per Capita
Argentina	$155	$8
Australia	$722	$106
Canada	$296	$245
Chile	$23	$79
France	$506	$353
New Zealand	$1,575	$186
United States	$1,780	$101

Source: Food and Agricultural Organization.

The relatively large size and export orientation of agriculture may also imply a relatively weak political position in terms of national development policy. There appears to be greater political support in New Zealand for small, import-competing sectors since these sectors were expected to provide significant employment opportunities for the unskilled segments of the labor force. Furthermore, incomes in rural areas have tended to exceed urban incomes — one of the very few countries in the world where this phenomenon has occurred. In short New Zealand has never experienced "The Farm Problem" to a large degree.

National economic policies have tended to be interventionist in some respects ever since the Great Depression. However, it is important to view these policies within the context of a small, dependent economy. The New Zealand economy was closely tied to the United Kingdom through trade, foreign borrowing, corporate linkages, and immigration until very recently. The linkages are still strong in the last three areas, even though exports to the UK have fallen to around 9 percent of total exports. In this environment, the terms of trade for major agricultural products were, at least to some extent, managed

TABLE 15.3 New Zealand's Agricultural Exports (million $NZ FOB)

	Year Ended 30 June 1991
Meat and Meat Products	$ 2,612
Dairy Products	$ 2,485
Wool	$ 1,044
Pastoral-Based Exports	$ 7,061
Fruit and Vegetables	$ 1,069
Agricultural Based Exports	$ 8,689
Total New Zealand Export of Goods	$ 15,147

Source: Ministry of Agriculture and Fisheries (1992).

by joint political forces. At their extreme during World War I and from 1939-53, export prices and volumes of major agricultural exports were administered by intergovernmental agreement.

In this environment of increased uncertainty regarding export receipts, New Zealand appears to have attempted to exploit an assumed "inelastic" export demand curve. The export demand was exploited not by taxing exports directly, but rather by imposing import protection. From 1938 onwards, the government imposed tight import restrictions (import licensing and high tariffs) on all competing goods. These tended to be final manufactures. The motive appears to have been to transfer income from the rural sector to the urban sector — a distributional rather than efficiency issue — particularly through the promotion of full employment (defined to mean unemployment no greater than 1 percent, Endres, 1984). That is to say, it was a policy of attempting to mobilize the agricultural surplus (rents) to finance industrial development. Infant industry arguments were also prevalent and in a broader sense of increasing the diversity of economic activities in a small country with limited natural resources (except for tourism!).

This antitrade bias in industry policy (which continues today, Janssen et al., 1991) constituted a tax on exports of varying pressure given the quantitative nature of the import protection, volatile world export prices and the inflationary nature of monetary policy over the period from 1938 to 1990. As a result, the government found it necessary to ease the pressure on the export sector (including agriculture) from time to time by compensating (tariff compensation) for the low, policy-induced internal terms of trade. This was done in a wide variety of ways, including subsidies, tax expenditures, regulations, state institutional development, and ownership and loan guarantees. One of the most explicit manifestations of this managed export tax approach occurred during the 1960s. Each year in the Budget speech, the Minister of Finance would review the income position of farmers. Then he would announce what subsidies (or even whole subsidy programs) would be added or deleted that year to ensure that farmers' incentives were adequate, but not "too" adequate. Seen as an income policy, it may have been relatively consistent, but as a series of input, output, and farm tax policies, it was quite ad hoc.

Economic policy was also interventionist over this extended period in terms of welfare-state provisions. This affected income tax rates and the performance of the labor markets. These policies probably did not affect the agricultural sector greatly because the effective tax rates of farmers were low given the special treatment afforded them as part of the tariff compensation package and because agricultural wage-setting was often a special case more attuned to the ability of the agricultural sector to pay. However, the presence and form of the extensive welfare state after 1975 was a major cause of fiscal deficits, which were then funded by money creation, and the resulting inflation did tend to affect primary agriculture adversely. The welfare state concept may have assisted agriculture in small ways through programs like price controls on bread, milk, and eggs. Originally designed as consumer subsidies, there is evidence that they were at least partially subverted and ended up benefitting producers instead. For example, wheat prices were set initially to help stabilize bread prices to consumers but ended up being set to stabilize the self-sufficency level, which assisted producers. Wheat prices over this period tended to be higher than potential import prices.

Definitions of Competitiveness

Hecksher and Ohlin took the theory of comparative advantage a step forward from the Ricardian version when they introduced factors of production and the production processes that transform them into goods and services. Competitiveness theory appears to push the framework forward another step. Competitiveness theory concerns itself with the processes that generate production factors and the organizational arrangements that combine them. It is a capital goods theory in a sense. The attention is focused heavily on human resource development and organizational theory at the individual, firm, and societal levels.

International competitiveness is taken here to mean the ability of an industry or subsector to profitably maintain or increase its market share over time. This connotes the idea of growth or size on the one hand and sustainability on the other. It also importantly implies that the resources used are rewarded appropriately. Competitive firms or industries are able to grow relative to their rivals, but who are their rivals? The appropriate comparison may depend on the objective of measuring competitive advantage. If the objective is to maximize the efficiency of global resource allocation, as it is in comparative advantage theory, then the comparison is with the global market shares of firms, including domestic and foreign shares. More precisely, it implies a measure of total factor productivity of the firm or subsector relative to all other similar firms in the world. This approach would parallel the measure of relative efficiency Leamer (1984) used in testing comparative advantage theory.

A definition of sustainable profitability creates a problem with agriculture because factor returns in excess of opportunity cost tend to be capitalized into fixed factor asset prices in the atomistic farm sector. Land prices might be used as an index of competitiveness, but there are few internationally comparable series available. There are also problems associated with separating market returns to land from the influences of subsidies by the state.

There is, however, another approach using quantity data, viz. the standard two-sector (exportables, importables) general equilibrium model with perfect competition. Real income or prosperity is maximized when the production possibility frontier has been moved to the greatest extent in a northeasterly direction with endogenous production factors like organizational skills and human resource development. This enables consumption to take place on the highest possible indifference curve. Industries that are internationally competitive in this economy are those that succeed in the international market — the exportable sector. Some exportable subsectors may be large and others small; some exportable subsectors will be growing, others shrinking. The degree of international competitiveness could denote the size of one exportable industry relative to another, or international competitiveness could be related to the relative growth rate of an exportable industry. The indices can be expressed in volume terms, if a small country is involved with exogenous terms of trade. That is to say, one may simply examine the composition of exports of the country in question.

However, the term international competitiveness also may be used to distinguish elements of competitiveness which are presumed to be different between domestic and foreign markets. In this context, a measure of international competitiveness would perhaps ignore domestic sales and focus on foreign sales alone. Here the focus is on shares of the world market, that is, inter-country sales alone. There is a parallel here between competitiveness on the global product side and Leamer's approach to defining relative abundance in global factor markets. This is an important distinction from the famous Leontief test of comparative advantage theory and is the second approach adopted in the next section.

Government policy intervention at home and abroad can affect international competitiveness in two ways. Both are very important in agriculture. The home country government can adopt import substitution policies, which effectively tax exports and reduce the size of the exportable sector, because the internal term of trade diverges from world price relatives. These policies may not be neutral in their effect on all exportable sectors. Agricultural barriers to trade imposed by foreign countries also cause New Zealand's terms of trade to deteriorate. Again, the effects may not be neutral across agricultural commodities. Both these effects have the potential to distort the true international competitiveness of particular agricultural subsectors through differential impacts on incentive levels and resource allocation.

Competitiveness of New Zealand Agriculture

Agriculture's Share of New Zealand Exports

New Zealand produces relatively large exportable surpluses of selected agricultural products — meat, dairy products, wool, kiwifruit, apples, and associated products. These relatively large surpluses have been produced for nearly 150 years as a result of high production levels relative to domestic demand. This is reflected in high agricultural self-sufficency ratios (Table 15.4). The self-sufficency ratio of 330 for beef means that New Zealand produces over three times as much beef as is consumed domestically.

Exports of processed and unprocessed agricultural products comprised at least 80 percent of total merchandise exports until the mid-1960s. Furthermore, the agricultural products were drawn from a narrow range of unprocessed wool, meat, and dairy products which had usually undergone only one transformation (manufacturing) stage (Table 15.5).

After 1966, two other trends developed. First exports of industrial products increased following the negotiation of a free trade agreement with Australia (NAFTA) and a currency realignment in 1967. Industrial product exports continued to increase during the 1970s and plateaued during the 1980s at around 45 percent of total exports. This is reflected in the other exports column in Table 15.5. The composition of agricultural exports also changed towards a greater degree of value-added and further stages of transformation. In this context, agricultural competitiveness appears to have improved through process deepening but has worsened relative to industrial sectors of the economy. Agricultural resource constraints may have tightened, especially for land and even labor in intensive farming like dairying. However, the declining share of agriculture in total exports is not likely to have been caused by Engel's Law in the New Zealand case because such a high proportion of its products are exported and the elasticity of export demand facing those agricultural products is very high (Findlayson et al., 1988).

One other element of international competitiveness may be the diversity of international markets. In this regard, New Zealand has improved its situation

TABLE 15.4 Selected Agricultural Self-Sufficiency Ratios for New Zealand

Commodity	Self Sufficency Ratio(%)
Beef	330
Veal	520
Mutton	280
Lamb	2,140
Pigmeat	100
Poultry meat	100
Butter	730
Cheese	430
Whole Milk Powder	630
Skim Milk Powder	360
Casein	1,240
Wool	650
Kiwifruit	500

Source: Adapted from Schroder (1987) and MAF (1992).

TABLE 15.5 Composition of New Zealand Exports by Value (percent)

Year	Wool	Meat	Dairy Products	Other
1950	40.9	15.7	29.8	13.6
1952	34.2	16.4	33.2	16.2
1954	36.6	20.6	27.3	15.8
1956	33.0	23.0	29.5	14.5
1958	32.0	29.4	23.3	17.6
1960	33.9	25.2	26.0	14.9
1964	36.8	24.4	22.7	16.1
1966	30.2	25.2	25.2	19.4
1968	19.3	31.2	25.4	24.1
1970	18.8	33.5	19.4	28.3
1972	16.6	28.6	25.6	29.2
1974	20.2	29.6	18.5	31.7
1976	19.1	24.6	16.5	39.8
1978	17.5	22.8	15.3	44.4
1980	18.1	23.0	15.4	43.5
1982	13.6	23.0	18.9	44.5
1984	12.9	20.0	16.3	50.8
1986	12.1	16.2	13.0	58.7

Source: Rayner and Lattimore (1991). Reprinted by permission.

considerably since 1970. Table 15.6 shows the share of New Zealand exports accounted by the top 20 countries in 1990. The final row of Table 15.6 gives the standard deviation of exports across the top 20 markets. This demonstrates the large increase in diversification from 1970 to 1980 and again from 1980 to 1990. In 1960, only nine of these markets took more than 1 percent of exports. In 1990, all 20 markets exceeded 1 percent. Australia was a rapidly growing market in all periods, quintupling its share from 1960 to 1990 to become New Zealand's largest export market. Japan has shown similar growth from a large base to become the second largest export destination. On the other side, the United Kingdom has reduced its share from 56 percent of exports in 1960 to 6 percent in 1990.

The newly industrializing countries, particularly in the Pacific Basin, have grown very rapidly in importance over the period. More traditional markets in Europe have grown more slowly or have shrunk. The United States remains prominent as the third largest export (and import source) destination, which is roughly the position it has held for 150 years.

TABLE 15.6 Country Shares of New Zealand Agricultural Exports

Country	1960	1970	1980	1990
Algeria	0.003	0.0009	0.006	1.097
Australia	3.727	8.0032	12.307	18.534
Belgium	1.772	1.9601	1.009	1.050
Canada	1.040	4.1604	1.902	1.464
China	0.848	0.3759	2.297	1.174
Fiji	0.391	0.8116	1.381	1.232
France	5.872	2.6432	2.459	1.059
Germany	3.301	2.7197	2.256	2.348
Hong Kong	0.133	0.4214	1.539	1.520
Indonesia	0.001	0.0718	1.171	1.052
Italy	2.046	2.1821	2.491	1.594
Japan	2.276	9.8314	12.329	16.425
Korea	0.001	0.1481	1.243	4.537
Malaysia	0.114	0.6682	1.128	2.489
Saudi Arabia	0.000	0.0098	0.695	1.144
Singapore	0.155	0.8032	1.440	1.429
Soviet Union	0.052	1.5578	4.870	1.150
Taiwan	0.002	0.2823	0.986	1.996
United Kingdom	56.412	35.4838	13.876	6.462
United States	14.662	15.2771	14.002	13.019
Standard Deviation	72	88	241	833

Source: MAF (1992).

International Competitiveness Index

As discussed previously, international competitiveness could also be defined relative to other exporters (firms, sectors, or countries). A static measure of competitiveness in this context is world market share — the proportion of intercountry trade exported by New Zealand. Table 15.7 provides estimates of the top 21 products in terms of world market share. That table shows that at this level of disaggregation, there are 10 products in which New Zealand has a greater than 10 percent world market share. The market shares of three products exceed 40 percent of world trade — kiwifruit, mutton and wool (when two categories are combined). Of the top 21 products, all but 5 are agricultural products (three of those are forest products), and all 21 are highly resource-based products.

A second measure of international competitiveness can be derived from world market shares. This index is simply the ratio of New Zealand's share of world exports in selected agricultural products to its share of world production

TABLE 15.7 World Export Shares of New Zealand Exports, 1987

Commodity	World Export Share
1. Kiwifruit	50.0
2. Sheepmeat	42.8
3. Wool, Scoured	36.4
4. Wood Pulp, Chemical	14.4
5. Sheep Pelts	14.3
6. Meatmeal, Offal	12.6
7. Buttermilk	11.6
8. Beef	10.6
9. Wool, Greasy	10.5
10. Butter	10.4
11. Sausage Casings	8.6
12. Casein	7.2
13. Edible Offal	6.3
14. Skim Milk Powder	6.0
15. Tallow	5.9
16. Fish Fillets	4.9
17. Apples	4.6
18. Wood Pulp, Mechanical	4.5
19. Race Horses	4.1
20. Plywood	3.5
21. Aluminium, Unwrought	3.3

Source: Derived from Crocombe, *et al.*, 1991.

of each of those products (Table 15.8). An index of 1.0 signifies that New Zealand exports the same proportion of world trade as it produces. For example, New Zealand is estimated to have had a competitiveness index of 5.0 for beef in 1989, which means that in 1989, New Zealand's share of world trade in beef was five times as great as its share of world production. The higher the index the more internationally competitive the country can be said to be because it is increasing its world market share relative to its production base.

The FAO production and trade data were used in these computations, and there may not be complete correspondence between the production and trade data (U.N.). Changes in inventories may also cause some interyear volatility in these indices. For all these reasons, these simple estimates need to be treated with caution.

The indices for New Zealand tend to be high in beef, mutton, apples, butter, and cheese. Furthermore, New Zealand indices of international competitiveness in its specialty products tend to be higher than indices of wheat and corn for countries which have a comparative advantage in grains (the index for the

TABLE 15.8 Index of International Competitiveness for Selected New Zealand Products

	Apples	Butter	Cheese	Nonfat Dry Milk	Beef	Mutton
1970	0.7	4.2	4.1	2.0	9.3	6.6
1975	4.6	2.8	3.5	4.2	3.8	4.9
1980	3.7	2.8	2.7	1.6	4.5	4.5
1985	6.1	3.6	3.5	1.3	5.6	3.4
1989	6.9	3.2	3.3	1.1	5.0	4.5

Source: Computed from FAO.

United States in corn and wheat are 3.1 and 3.0, respectively, in 1989). If one takes beef, mutton, and apples as examples, this latter comparison says that New Zealand's degree of comparative advantage in these products is greater than the degree of comparative advantage of the United States in grains. Accordingly, this index can be used to provide some cross-country and cross-commodity comparisons of competitive advantage.

The indices in New Zealand tend to rise over the period for beef, apples, and butter, providing some indication the international competitiveness in these products is rising. The indices for mutton, nonfat dry milk (NFDM), and cheese tend to be static or falling over the period, perhaps indicating a weakening of competitiveness. The New Zealand index is lower than Chile's index for apples (which was 8.1 in 1989) and lower than Canada and Australia for NFDM (3.2 and 1.4 respectively in 1989). This may indicate some opportunities to improve the New Zealand subsectors.

Sources of Competitiveness

The PR discussed earlier attributed much of New Zealand's agricultural production potential to natural resource endowments. This is misleading from a technological perspective. New Zealand soils are not naturally fertile and the topography is very uneven, for example. In agriculture, natural resources are necessary but not sufficient for competitiveness. Research and development have played a crucial role in New Zealand as they have elsewhere.

One important agricultural system is based on the intensive grass feeding of highly developed C5 grass species, in situ, to unhoused livestock. Farm grazing systems, especially on hills, have been developed on this basic technology enabling relatively high labor productivity on pastoral farms.

Some regions of New Zealand are not only temperate but have a long growing season for plants with a very even distribution of sunshine hours. This is ideal for the development of a wide range (maturities) of apple, pear, and related crop varieties. These conditions also exist in parts of Chile but not in most other countries.

There are a number of aspects to the success that evolves from these attributes. First, there are few social barriers between producers and researchers. In this environment, ideas and innovations discovered by farmers can be relatively made known to researchers for further development before being transferred to a whole agricultural subsector. In a sense, the whole subsector becomes a national experimental farm. For example, the gala and braeburn strains of apples were both discovered accidentally by growers in New Zealand. Researchers were quickly appraised of their existence. The research and technology transfer system was accordingly able to quickly refine the breakthroughs and create technological packages for general distribution. In short, the creation, refinement and adoption of new agricultural technology may be aided by a social and agricultural system that involves complex feedback loops (McArthur, 1983). A linear, top-down technology generation system may be less efficient. Second, continued competitiveness involves having a system in place which ensures that the process of grower/researcher interaction continues. New Zealand appears to have been very good at that.

On the production side, this advantage grew out of large investments in land development and infrastructure in and for farming, and a culture emphasising a farming ethic, farming skill development, access to land and specialized technological development for more than 150 years. Agricultural development was export-market lead from the early 1800s in typical colonial fashion. Australia, the United States, and the United Kingdom were all important markets initially. However, after about 1870, export demand was dominated by a single market—the United Kingdom. The advent of refrigeration was very important in this switch, creating an effective demand for meat and dairy products. Increasing quantities of land suitable for producing these products became available at the same time as a result of battles and wars between the Maori tribes and the most recent colonizing people.

The international competitiveness of the agricultural marketing system is difficult to assess and quite controversial. This is because most major agricultural exports are under the direct or indirect control of statutory, producer-controlled marketing boards. These institutions are a central focus for agricultural policy in addition to their role as marketing institutions. They often have monopoly export rights (sometimes import rights as well) over their commodity. Some of these institutions are large and growing multinational organizations, often involved in diverse business activities well beyond the marketing of the product for which they are legally responsible.

Since the larger marketing boards have existed for at least 50 years, it has been very difficult to carry out counterfactual analysis, so there is little hard evidence of their marketing performance. There has been little political incentive to change agricultural marketing policy; indeed, the reforming fourth Labour

Government (1984-90) did not tamper with marketing boards and even created a new one for kiwifruit. Nevertheless there is some evidence that underinvestment in marketing is occurring in New Zealand as a result of marketing board legislation. It was strands of such evidence that presumably stimulated the quotation earlier from the PR, although it may well have been made in response to political economy studies of agricultural institutions which are extremely critical of the structure, conduct and performance of marketing boards in Australia (e.g., Sieper, 1982 and Watson, 1983).

The CEO of one of New Zealand's fastest growing meat processing companies, which is renowned for its vision and productivity growth, recently said, "The [New Zealand] Meat [Producers] Board has been the leading influence in reducing access [to foreign markets]," Thompson (1992, p.5), [author's parenthesis]. Here, Thompson is referring to regulations and orderly marketing arrangements imposed on exporters to combat "weak selling." This raises a question as to whether a marketing board will tend to act as a subsidy or a tax on exports. In their evaluation of agricultural marketing board performance in New Zealand, Zwart and Moore (1990,p.261) concluded "The empirical evidence presented here, although scant, does little to support the observation that these institutions have lead to increased market returns for New Zealand."

Policy Distortions

As explained, New Zealand has pursued an import substitution strategy in the past. Since most of its agriculture sector is export-oriented, this has meant that the internal terms of trade have been moved against agriculture, reducing its size and incentive to produce. This might be interpreted as a policy-induced reduction in competitiveness. The size of this bias in exportable/importable industry incentives has been estimated using the Clements and Sjaastad framework. Their framework attempts to measure the export tax (so-called true tax) which is equivalent to the net effect of import protection and export subsidization policy. It is estimated that the true export tax facing agriculture over the period 1981 to 1989 averaged 12 percent, but has fallen to around 8 percent in 1992, Duncan et al. (1992). This means that the output and marketing performance of New Zealand agriculture is a response to a domestic market environment which has been negatively affected by policy factors to a significant degree. In a neutral policy environment, more resources could be expected to be devoted to agriculture, improving the sector's competitiveness. Competitiveness might also be raised in the global sense if there are economies of scale in processing, product development, and marketing.

Of even more importance is the effect which foreign agricultural policy has on the competitiveness of New Zealand agriculture by limiting export access through tight import restrictions and export subsidies. The extent of this intervention relative to New Zealand is illustrated in Table 15.9, which shows

TABLE 15.9 Degree of Agricultural Subsidization in Selected Countries (Producer Subsidy Equivalent).

Country	1979-86 Average	1991
Australia	12%	15%
Canada	32%	45%
EC	36%	48%
Japan	67%	67%
New Zealand	23%	4%
U.S.A.	28%	29%

Source: MAF (1992).

the percentage Producer Subsidy Equivalent (PSE) for 6 OECD countries for two periods. New Zealand has traditionally had a very low level of agricultural subsidization, on the order of the 1991 PSE. Over the period 1979-86, it followed a much stronger tariff compensation policy, resulting in the aberrant PSE for that period.

The higher PSE's for the representative countries shown in Table 15.9 cause New Zealand exports to be lower, and its price performance internationally to be poorer. This negatively affects international competitiveness of New Zealand agriculture. One way of measuring this effect indirectly is to examine the relative responsiveness of countries in the event that agricultural subsidies were to be removed. If dairy subsidies in the OECD countries were to be reduced by 10 percent (and dairy quotas decreased by 10 percent), milk production in all OECD countries would decrease, except in Australia where production would increase by 3 percent and in New Zealand where the increase would be 17 percent (OECD). This reflects a very high level of international competitiveness for the New Zealand dairy subsector. Similar results could be expected for New Zealand beef, mutton, apples, kiwifruit, and other major agricultural exports, although New Zealand may not always be the most competitive country.

References

Cartwright, W. "International Competitiveness of Land-Based Industries: Approaches to Research." Discussion Paper, No. 131, Agribusiness and Economic Research Unit, Lincoln University, 1991.

Crocombe, G., M. Enright, and M. Porter. *Upgrading New Zealand's Competitive Advantage.* Auckland: Oxford University Press, 1991.

Duncan, I., R. Lattimore, and A. Bollard. "Dismantling the Barriers: Tariff Policy in New Zealand." Research Monograph 57. Wellington: New Zealand Institute of Economic Research, 1992.

Endres, A. "The New Zealand Full Employment Goal." *New Zealand Journal of Industrial Relations* 9(1984): 33-44.

Findlayson, J., R. Lattimore, and B. Ward. "New Zealand's Price Elasticity of Export Demand Revisited." *New Zealand Economic Papers* 22(1988): 25-34.

Holmes, G. *Revolution in Agriculture.* London: Todd Publishing Co., 1946.

Janssen, J., G. Scobie, and J. Gibson. 1991. "Liberalisation in the New Zealand Economy: Reforms, Consequences and Lessons." Department of Economics, University of Waikato (Mimeographed), 1991.

Leamer, E. *Sources of International Comparative Advantage: Theory and Evidence.* Cambridge, MA: The MIT Press, 1984.

Ministry of Agriculture and Fisheries. "Situation and Outlook for New Zealand Agriculture." Wellington: MAF, 1992.

McArthur, A. "Transfer of Technology, AGPOL '83." Wellington: New Zealand Institute of Agricultural Science, 1983.

Organization for Economic Cooperation and Development. *Changes in Cereals and Dairy Policies in OECD Countries.* Paris: OECD, 1991.

Porter, M. *The Competitive Advantage of Nations.* New York: The Free Press, 1990.

Rayner, A., and R. Lattimore. "New Zealand." In *Liberalising Foreign Trade,* Vol. 6, edited by D. Papageorgiou, M. Michaely, and A. Choksi. Oxford: Basil Blackwell, 1991.

Sieper, E. "Rationalising Rustic Regulation." N.S.W., Australia: Centre for Independent Studies, 1982.

Thompson, G. "Alumni Dinner Address." Department of Economics and Marketing, Lincoln University, (mimeographed), 1992.

United Nations. "Production and Trade Yearbook." Rome: FAO.

Watson, A. "The Australian Wheat Board Marketing Agency or Plaything for Politicians, Public Servants and Farm Organisations." N.S.W., Australia: Centre for Independent Studies, 1983.

Zwart, A., and W. Moore. "Marketing and Processing." In *Farming Without Subsidies,* ch. 13, R. Sandrey and R. Reynolds. Wellington: GP Books, 1990.

16

Assessing the International Competitiveness of the Danish Food Sector

Aage Walter-Jørgensen

Danish agriculture is an important supplier of agricultural products to world markets. The main part of the export is animal products originating from domestic production and traded, for a large part, by cooperative processing and marketing firms owned and controlled by farmers. Primary agriculture thus holds a central position in the Danish food sector as farmers have the ultimate responsibility for the management of their cooperative associations. The competitive position of the food sector, therefore, strongly depends on the performance of primary agriculture and of the ability of farmers to adapt to changes in the economic and political environment.

This chapter investigates the development of Danish farm production and exports, emphasizing major factors behind the present position of the Danish food sector. Emphasis is given, in particular, to the development in productivity, organization of processing and marketing and to the impact of agricultural policies on the competitive situation of the Danish food sector.

Export Performance of Danish Agriculture

Danish agriculture provides the basis for large-scale production and export of farm products. In total, the agro-industrial complex in 1990 contributed about 8 percent ($9 billion) to the national product, 45 percent of which derived from primary agriculture, about 20 percent from processing and marketing of farm products, and about 35 percent from the farm supply industry (Table 16.1). The production of cattle and milk provides the largest contribution to value added, followed by pig production and cropping. Three-quarters of the value added in the farm sector is associated with animal production.

TABLE 16.1 Gross Factor Income in the Danish Agro-Industrial Complex, 1990

Billion $U.S.	Primary Agriculture	Processing Industries	Supply Industries	Total
Crop products	1.3	.2	0.9	2.4
Cattle	2.0	0.6	1.1	3.7
Pigs	0.7	1.1	1.0	2.8
Poultry	0.1	0.1	0.1	0.2
Total[1]	4.0	1.9	3.1	9.1
Per cent of GDP	3.6	1.7	2.8	8.1

[1]Excluding pelt animals.

Source: Institute of Agricultural Economics.

About 67 percent of the Danish farm production is exported. Exports account for about 60 percent of the production of butter and for 70-75 percent of the production of cheese, beef and veal and pork. Denmark imports considerable quantities of feedstuffs (in particular protein feedstuffs) but is a net exporter of cereals. Including canned meat and canned milk products, agricultural exports account for about one-fifth of the country's total export value.

Danish exports of agricultural products represent about 2.5 percent of international trade in these products. In the case of butter and cheese in 1990, Denmark held 6 and 9 percent, respectively, of global exports, whereas for fresh and frozen pork and bacon, the share of global trade amounted to 20-30 percent (Table 16.2). Denmark is leading in the production of pelts (mink) with 35-40

TABLE 16.2 Denmark's Share of Agricultural Exports, 1990[1]

	Million $U.S.	Percent of Global Export
Agricultural products, total	8,290	2.6
Cereals	517	1.4
Sugar	136	1.0
Butter	187	5.9
Cheese	693	8.6
Beef and veal	418	3.1
Pigmeat, total	2,061	22.8
Bacon, ham etc.	439	30.6
Fresh frozen pigmeat	1,622	21.3
Poultrymeat	112	2.8
Pelts (mink)[2]	10	35-40
Pot-plants[3]	308	25-30

[1]Including intra-EC trade.
[2]Million of pelts produced.
[3]Exports to main European Markets.

Source: FAO, Trade Yearbook and national statistics.

TABLE 16.3 Danish Agricultural Exports and Their Major Destinations, 1991[1]

	Billion $U.S.	Per cent			
		EC-12	USA	Japan	Others
Animal products	3.9	64	4	16	16
Pigmeat	2.2	64	6	26	4
Butter	0.2	83	-	-	17
Cheese	0.7	53	4	7	36
Vegetable products	1.4	76	-	-	24
Canned meat	0.7	53	18	2	27
Canned milk products	0.3	17	-	-	83
Pelts	0.4	35	4	-	61
Total	6.7	61	5	10	24

[1]The figures are exclusive of export restitutions.

Source: Denmark Statistics.

percent of world production and Denmark is a large supplier of pot-plants to the European market.

Judging from this export pattern, it appears that Danish agriculture is holding its strongest export position in products for which government support is minimal (e.g., pork, pelts and pot-plants). Indeed, Danish exporters seem to concentrate on a relatively small number of products (focused differentiation) for which there is a large market. To promote exports, a considerable part of the produce is sold under the "Danish" mark, although market research carried out among European consumers shows limited recognition of Danish products at retail levels (Ministry of Agriculture, 1991). One explanation might be that, in the case of meat, a large share of exports consist of semi-processed products.

About 55 percent of the Danish farm exports (including export restitutions) was sold on EC markets in 1991. The largest export item is pork, accounting for a total value (excluding restitutions) of $2.2 billion in 1991 (Table 16.3). Exports of canned meat, most of which is pork, amounted to $0.7 billion. Next to the EC, Japan is the largest market for fresh and frozen pork, whereas the United States is the main importer of Danish canned meat outside the EC. The main outlet for Danish butter is the UK market, whereas cheese is exported to a number of countries. The largest export item among cheese products is Feta-cheese for the Middle East. Danish exports to countries outside the EC has increased both in absolute and relative terms since Denmark became member of the EC in 1973.

As the EC price of most farm products is higher than the world market price, exports of such products to countries outside the EC is not feasible without government support. In 1990/91, export restitutions granted for Danish exports to third countries amounted to 29 percent of the value of that export, compared with 18 percent in 1973-74. By far the largest support is paid to dairy products but also exports of pork to the United States and Japan depend, to a varying degree, on export restitutions. The EC provides no price guarantees for grain-based products (pork, poultrymeat and eggs), but export restitutions are granted as compensation for higher feeding costs due to cereal prices being held above world market prices. As the granting of export restitutions depends on market conditions, export subsidies tend to vary over time. Due to Denmark's special veterinary status, most export restitutions granted to pork in the EC go to Danish exports.

Structure of the Agro-Industrial Complex

The export position of Danish agriculture can be traced back to the latter part of last century where, in the wake of the industrial development of the 18th century, there was a growing demand for processed agricultural products. This, combined with an increasing supply of cereals from the American market, provided the basis for a rapid expansion of animal production. It was also of importance that Denmark, contrary to many other European countries, chose a liberal market policy which allowed Danish producers to take advantage of low world market prices of cereals.

The expansion of animal production was supported by establishment of cooperative processing and marketing associations which still dominate the Danish food sector. The co-operatives are based on the principle of "one man, one vote," which has given rise to criticism from "large" producers. This problem seems, however, gradually to have vanished as primary production is being concentrated on larger and more specialized farm units. It might also have affected market structures that cooperative associations are subject to certain tax concessions regarding capital accumulation.

The marketing of pork is managed almost exclusively by five cooperative slaughter houses, compared with 18 cooperative and two private slaughter houses in 1980 (Table 16.4). Exports of pork are handled by individual houses or by the Danish Bacon Factories' Export Association, which is organized as a commercial company trading fresh and frozen pork.

About 90 percent of the dairy products is handled by cooperative dairy associations. The number of dairies has fallen considerably over the last few decades and the remaining dairy associations are amalgamated into larger co-operatives, of which the three largest in 1991 had control of 84 percent of total milk deliveries. The private dairies are relatively small and primarily orientated towards home market sales. Exports of butter and cheese are co-ordinated through the Danish Butter Export Board and the Danish Cheese Export Board, respectively, but physically handled by individual dairies or by the Cooperative Butter Export Association.

The dominant position of cooperative associations means that a large part of the agro-industrial complex is controlled by a confined group of animal

TABLE 16.4 Structure of Processing Industries in Denmark

	1980	1991
Pig slaughtering companies		
Cooperatives	18	5
Private	2	0
Slaughtering units, total	36	25
Dairies		
Cooperatives	147	26
Private	61	27
Share of milk production, per cent		
Cooperatives	88	92
Private	12	8

Source: The Federation of Danish Pig Producers and Slaughterhouses and Danish Dairy Board.

producers. More than 95 percent of the Danish farms are owner-operated family farms. Less than half of the 76,400 farms with more than 5 hectares in 1990/91 were full-time farms (i.e., farms with at least one man-year of input). Of these full-time farms, 18,600 were classified as dairy producers (accounting for 91 percent of total milk production), whereas the 9,100 full- time pig producers represented about 75 percent of pig production (Table 16.4). Part-time farmers mainly concentrate on crop production.

The International Competitive Situation

Denmark has increased its share of pig production in the EC during the 1980s but lost market share in dairy and beef production. In the poultry subsector, Danish producers have maintained their share of production. During the 1980s, pig production in Denmark increased by one-third as compared with the 15 percent increase in the EC taken as a whole (Table 16.5). Pig production in

TABLE 16.5 Change in Animal Production for Selected EC Countries, 1978-80 to 1988-90 (Percent)

	Milk	Beef and veal	Pork	Poultry	Eggs
Denmark	-9.5	-14.8	32.7	28.5	8.0
W. Germany	-2.1	10.1	2.4	17.9	-14.7
France	1.2	-1.3	-0.8	48.4	12.3
Netherlands	-3.0	26.3	65.3	41.5	34.4
UK	-5.9	-8.0	6.0	42.2	-14.4
EC-10	-0.3	3.4	15.1	33.5	-1.3

Source: Commission of the European Communities.

Netherlands, however, exceeds that of Denmark both in size and rate of growth, with a 65 percent increase during the same period. In other EC countries, the rate of increase in pig production was below average, meaning that those countries have lost market share to Netherlands and Denmark.

Milk production in Denmark fell by nearly 10 percent during the 1980s. It should be noted that milk production in the EC became subject to quotas in 1984, when milk production was reduced and the distribution of milk production frozen among member countries. Yet, the main part of the fall in Danish milk production took place before introduction of the milk quota, indicating a certain lack of competitiveness in the Danish dairy subsector (Hansen, 1985). The same conclusion applies to beef and veal, where countries like Netherlands, Ireland and Germany have a comparative advantage due to a high percentage of permanent grass land.

The rate of increase in Danish poultry production has been close to the average of the EC, whereas Netherlands, UK and France, in particular, have increased their market shares considerably. The egg subsector in Denmark, whose production fell by more than half during the preceding decades, improved its market position during the 1980s.

The high intensity of animal production sets focus on the impact of agriculture on the environment, pollution of streams and open waters being of primary concern in Denmark. In order to limit the leaching of nitrogen from agriculture, herd sizes are limited, as is the disposal of slurry. The political pressure for such restrictions varies from country to country and further strengthening of environmental regulations could have considerable impact on the international competitive position of agriculture.

Wage Competitiveness Improved During the 1980s

The competitive position of Danish agriculture is influenced by internal as well as external factors. Being a member of the EC, Danish agriculture is subject to regulations of the Common Agricultural Policy (CAP) which provides an outlet for the farm produce at the common EC prices. Yet, marketing conditions and differences in product quality affect price relationships among member countries. In an analysis of the development in EC terms of trade, Walter-Jørgensen found that price conditions vary considerably among countries and products. Thus, the rapid growth of production in Netherlands is coinciding with rather advantageous price conditions during the 1980s, whereas Denmark has had less favorable price conditions, particularly in the case of poultry.

The competitiveness of the food sector is also affected by labor and capital costs. Denmark has a high wage level; yet, in recent years, wages have increased more slowly than in other countries (Figure 16.1). Exchange rate adjustments, which can cause considerable fluctuation in terms of trade from year to year,

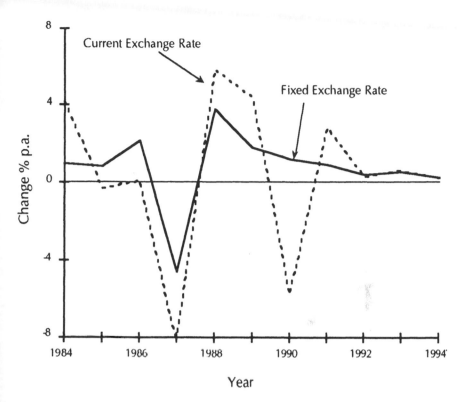

FIGURE 16.1 Wage Competitiveness in Denmark. *Nota:* Wage competitiveness is measured by the ratio of wage increases in Denmark relative to wage increases abroad weighted by Danish industrial exports. *Source:* Economic Council, 1992.

do not alter the general picture that the Danish food industry has benefitted from increasing wage competitiveness during the 1980s.

The cost of capital is affected by taxation and inflation. Denmark typically has high nominal interest rates but, when adjusted for inflation, the difference in cost of capital as compared with neighboring countries has gradually diminished during the 1980s. As an example, the real rate of interest (adjusted by consumer prices) in Denmark was about 3 percentage points higher than in West Germany in 1980, falling to about 1 percent by the end of the decade. Because of changing tax rules, however, real capital costs to Danish firms have increased during the same period.

Productivity Is Crucial for Competitiveness

Increasing productivity in primary agriculture seems to be a major factor behind the growth in Danish food exports. In an analysis of productivity changes in agriculture, Larsen et.al. found that labor productivity in Danish agriculture has increased by 6-7 percent annually during the 1970s and 1980s as compared with 4-5 percent in other EC countries. It is, however, in pig production that productivity has increased most, whereas the growth in productivity in cropping and dairying has been somewhat more moderate (Hanson, 1990).

The high rate of productivity increase in the pig subsector can be traced to the middle of the 1970s where introduction of cross breeding and new production systems laid the groundwork for a rapid growth in the number of pigs produced per sow (Figure 16.2). Until that time, pig production was primarily based on the Danish Land Race, which was fit for bacon production but troubled by low fertility. As the crossing technique was widely applied abroad, the technology gap closed. The latter is supported by the observation that the rate of increase in productivity in Danish pig production has fallen from the 1970s to the 1980s, when it averaged 3.3 percent (Hanson, 1990).

Productivity seems to have increased somewhat more in the 1980s than in the 1970s for other lines of production. In recent years, productivity in the dairy subsector has depended heavily on an administrative re-allocation of milk

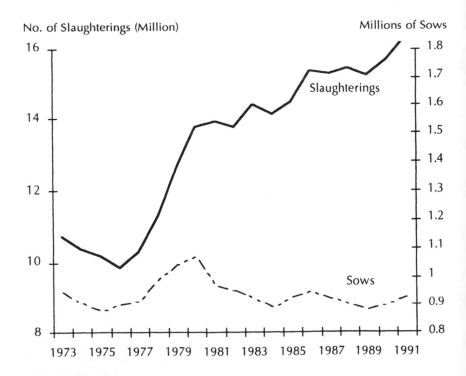

FIGURE 16.2 Herd of Sows and Pig Production, Denmark.

TABLE 16.6 Producer Subsidy Equivalents, Average 1989-91[1]

Percent	EC	USA
Wheat	44.4	39.9
Coarse Grains	47.0	25.0
Milk	65.8	58.4
Beef and veal	55.0	31.3
Pork	8.0	6.3
Poultry	24.2	10.0

[1]Net percentage PSE, weighted average.
Source: OECD.

quotas. There is little doubt, however, that the freezing of production shares among member-countries has reduced productivity improvements in the dairy subsector.

Limited Effects of EC Reforms

The market position of the Danish food sector is supported by the CAP. Without this support vegetable, milk, and beef production would have been considerably lower in Denmark. The production of pork, poultrymeat and eggs is less dependent on CAP price policy. The EC provides higher support for its agriculture than does the United States. Yet, the variation in support among products is very much the same (Table 16.6). The highest support is paid to milk and beef producers, followed by cereal producers. The support for other livestock producers is mainly structural support as there are no price guarantees for pork, poultrymeat, and eggs in the EC. The internal market for these products is protected by a system of import levies and exports restitutions which, as mentioned before, are linked to the price of cereals. A reduction of price support will therefore reduce the price of these products but leave the competitive position unaffected.

The reform of the CAP, which was passed by the Council of Ministers in 1992, will have limited effect on the Danish food sector. The most important change is related to vegetable production where the price of cereals is to be lowered by almost one-third over three years. On average, however, the total level of support is maintained through acreage-linked payments. Farmers must set aside 15 percent of their crop acreage in return for acreage payments on planted and fallowed land. The reform is a rather mild solution for the cereal subsector compared with the present line of policy, which has involved a fall in the real price of cereals of 6-7 percent per year since the mid-1980s.

The EC reform involves few changes for the dairy and beef subsectors. Milk quotas will be reduced by 2 percent and the price of milk by 2.5 percent over

TABLE 16.7 Effect of EC-Reform on Danish Agricultural Production and Trade

	Change in EC-price percent	Percent change in:		Exports, million tons	
		Production	Consumption	Before	After
Cereals	-28	-9[a]	1	1,574	808
Milk	-1	-2	-	2,632[b]	2,535[b]
Beef	-15	-9	6	137	111
Pigmeat	-10	2	4	822	828
Poultrymeat	-13	-5	5	59	50

Note: The analysis shows the effect over 3 years of 28 percent price reduction for cereals, 15 percent for beef and a derived fall in the prices of pigmeat and poultrymeat corresponding to the fall in costs of feedgrain. The basis for calculations are 1986/89 data for production, consumption and prices (CRONOS-data) and medium term supply and demand elasticities (OECD-data), including cross- price elasticities for meat demand.
[a]Production is expected to fall less than set-aside due to "slippage."
[b]Milk equivalents.

Source: MITCAP-model, Institute of Agricultural Economics.

three years, the latter being modified by elimination of the existing 1.5 percent co- responsibility levy. Thus, the reform is seen as a continuation of the present policy where milk quotas determine the supply of milk and the distribution of production among member-countries. The price of beef is to be reduced by 15 percent over 3 years, but producers of beef cattle will receive headage payments in compensation. These payments will probably be of little benefit to Danish producers because they are primarily directed towards lower income areas.

The production of pork, poultry, and eggs will be slightly affected by the reform. Indeed, the lowering of cereal prices will reduce production costs but, as mentioned before, lower cereal prices tend to be balanced by lower product prices, the effect on supply therefore being diminutive. Due to lower meat prices, consumption in the EC is predicted to increase, thus making agriculture less dependent on meat exports. Lower cereal prices will also make Danish exports to third countries less dependent on EC export restitutions.

The impact of the EC reforms on Danish agriculture is illustrated in Table 16.7, which shows the effect on production, consumption and exports.[1] As indicated by the figures, the production of cereals and beef is expected to fall considerably, whereas milk production is expected to fall by merely 2 percent. Pig production might increase slightly as producers will benefit from a relatively large fall in feed costs.[2] Because of lower prices to consumers, consumption will increase, thus reducing export potentials for milk, beef and poultrymeat, whereas little change is foreseen in the case of Danish exports of pork. Such results should, of course, be treated with great care as they are subject to considerable uncertainty. Yet, the analysis supports the observation that the EC reform will probably have limited impact on the export performance of the Danish food sector.

Concluding Remarks

As indicated above, the Danish food sector seems to maintain its strongest position in areas where market forces are prevailing. An important example is production of pork where Denmark provides considerable export volumes to world markets. Historically, this can be traced to the end of the last century when Denmark chose a free market policy for its farming industry, allowing for a growing animal subsector based on world market prices for cereals. It is also important to note that Denmark quite early succeeded in organizing an export-oriented food industry, which has been the platform for further development up to today.

Success in international trade is dependent on productivity and economies in production which ultimately are determined by the ability to utilize existing knowledge and new technology. Compared with other EC countries, it appears that labor productivity in Danish agriculture has improved considerably over the last two decades. This is particularly true for the pig subsector, where the closing of a "technological gap" has provided the base for expansion in production and exports. It may, therefore, require considerable efforts to maintain a competitive advantage in this subsector. Productivity has increased somewhat less in the dairy subsector and, considering the constraints on production in the EC, the probability of further expansion in this field is rather small.

The Danish food industry tends to concentrate on a few, specialized products of high quality for which there is an international market. This may have affected the market position of the industry. The dominant cooperative structure of the food sector might have furthered this market strategy, but the organizational form is probably of little importance to the economic performance of the food sector.

The frame of policy for agriculture is crucial for international competitiveness. One should not underestimate the importance of having the EC, accounting for about 340 million inhabitants, as a domestic market for Danish farm products ensuring outlets for surplus products in case of falling exports to third countries. For several products, such as beef and dairy products, exports are dependent upon support from the EC. It is, however, in subsectors exposed most to competition (pork, poultry, pelts etc.) that the Danish food industry has its strongest potential for expansion in the coming years.

Notes

1. The results are based on a static, comparative analysis showing the full effect of the EC-reform over three years. A partial equilibrium model is used to calculate distributional effects of the CAP, using CRONOS data and OECD price elasticities.

2. Due to differences in feed composition, the reduction in cereal prices will affect production in member countries differently. For instance, in Holland, pig production is based to a large extent on cereal substitutes imported at world market prices, whereas in Denmark pig production is mainly based on home-grown cereals. A lower price of cereals in the EC tends, therefore, to benefit Danish producers more than Dutch producers.

References

Commission of the European Communities. *The Agricultural Situation in the Community.* Luxembourg: EEC, 1991.

Economic Council. *Danish Economy* (May 1992).

Hansen, J. *The competitiveness of Danish Dairy Production* Institute of Agricultural Economics Report No. 23, 1985.

Hansen, J. *Productivity and Terms of Trade in Danish Agriculture, 1973/74-1987/88.* Institute of Agricultural Economics, Report No. 45, 1990.

Institute of Agricultural Economics. *Danish Agricultural Economy,* 1991.

Ministry of Agriculture. *International Marketing of Agricultural Products from Danish Food Suppliers.* København: Ministry of Agriculture, 1991.

Walter-Jørgensen, A. 1992. Animal Production in Danish Agriculture. Tidsskrift for Landøkonomi. (Forthcoming).

PART FIVE

Commentary and Implications

17

Commentary and Implications

Commentary on the Concept of Competitiveness

T. Kelley White

Commenting on the concept of competitiveness has been an interesting exercise for me. During the mid 1980's I was involved in an Economic Research Service effort to determine whether U.S. agriculture had lost it's competitiveness, and if so, why, and, what could be done about it. Before we could provide definite answers I moved to FAO and became absorbed by a different bureaucracy and development problems rather than competitiveness problems.

I had not thought about competitiveness as such for almost two years until Maury Bredahl suggested that thinking on the subject had progressed and that, if I wanted to be competitive in this role, I should read some of the current literature. A little background reading, and reading the papers prepared for this book, has been a little like returning to your hometown after a long absence to see if it is as you remember it. I have just arrived back in the town called "Competitiveness" and while I haven't had an opportunity to visit many places, things are more like I remembered them than I had hoped.

There is still confusion about: what competitiveness is; what its determinants are; what our various bodies of received theory contribute, whether there is, or, can be, a generalized theory of competitiveness; whether government and policy are cause, effect or both; how to measure it; and about the appropriate level of economic aggregation at which competitiveness can be observed and influenced. Abbott and Bredahl do a masterful job of identifying the problems with the concept of competitiveness as employed, of pointing to many of the sources of these problems and of identifying the "right questions." But, then they missed the opportunity to suggest that the terms "competitiveness" and "competitiveness problem" be knighted and appointed to the House of Lords where they would have status but cause relatively few problems. Economists and policy makers (and maybe managers) could them get on with seeking and applying answers to the right questions. Instead, they proceed to identify the determinants of competitiveness, propose a conceptual framework for agricultural competitiveness and issues in agricultural competition. This after having quoted Robert Reich as having said that "rarely has a term in public discourse gone so directly from obscurity to meaninglessness without an intervening period of coherence (Wall Street Journal)."

McCorriston and Sheldon in their paper say "... it is difficult to find a useful definition of 'competitiveness'." Then they picked one and proceeded to identify aspects of the competitiveness problem for which traditional trade theory is inadequate and for which recent developments in theory are more adequate.

Van Duren, Martin and Westgren say "the use of the term competitiveness has become so common place that it risks becoming meaningless." They then propose a framework for assessing national competitiveness and the role of private strategy and public policy. Lau, to his credit, in a paper treating processes central to most definitions of competitiveness, managed to avoid use of the term until the penultimate paragraph.

The problem with competitiveness as a term describing the condition of an economic unit (firm, industry or country) is not only that it means different things to different people. Equally as serious is that, at whatever level of aggregation competitiveness is defined, its determinants are nearly infinite. In most discussions of competitiveness the answer to the question "What affects competitiveness of this unit" is "everything under control of the unit and constituting its environment." Thus, while competitiveness of economic units is important and of interest to economists, policy makers, managers and others, it is not a very useful indicator on which to focus to either identify problems or prescribe solutions.

Healthfulness and Competitiveness

To economists the word "competitiveness" has such a connotation of "goodness" that it is difficult to accept that it isn't a very useful indicator of the condition of an economic unit. It may be helpful to draw a few analogies with another field: human health.

Health refers to the overall condition of the complex system comprising a person. When all the subsystems are functioning normally we would usually say the person is "healthy". We also speak of healthy subsystems of the body as well as healthy populations of people. Thus, saying that a firm, industry or nation is competitive is a statement about its general condition or like saying a person is healthy. Likewise to say an economic being is uncompetitive or has a competitiveness problem is like saying I have a health problem, I am unhealthy or I am sick. It may get you sympathy, and a suggestion to go to the doctor for a checkup. Competitiveness in this sense, like healthiness is not directly observable but is deduced from various observable indicators. And, in both cases it is usually the lack of competitiveness or healthiness that attracts our attention.

If we go to the doctor with an unhealthiness problem we probably don't call it that, but given some of the symptoms, the doctor will check various indicators for malfunction of subsystems, ask about abusive activities, run tests and other wise attempt to isolate the specific problem and its cause. Then a prescription is made to treat the specific problem. There is not one but many causes of unhealthiness and not one prescription for returning to a healthy state but many specific to the disease, illness or injury. Using competitiveness in this sense makes clear the folly of seeking a cure for the "competitiveness problem".

While there is no disease called unhealthiness and no general treatment for unhealthiness, there are rules of behavior, which are prescribed as conducive of healthiness. We can all remember from health class rules such as drink eight glasses of water per day, exercise daily, eat a balanced diet (although we may later learn some of those good foods are bad for us), don't smoke, don't drink too much and etc. Even though such general rules of behavior exist and are useful for staying healthy, we don't confuse them with treatment of specific diseases and run for the old health books when we break a leg or get cancer. It may be useful to look for general rules of behavior that are conducive to competitiveness but these should not be confused with treatment of specific problems.

How should we, as economists (the medical profession for the economy), change the manner in which we interact with policy makers, business people, and the general public in order to reduce confusion and put competitiveness in this suggested context? What do we as a profession need to do? Providing answers to these questions is beyond the scope of this paper and the area of competitiveness of this discussant. But maybe a few thoughts on some of the things needed will help generate discussion.

Staying with our analogy, I would propose the following:

1) Define what is meant by healthy (competitive).
2) Identify indicators of the state of healthiness.
3) Identify diseases (and other conditions) that negatively affect health and identify treatments specific to diseases.
4) Be on the look out for changes in the environment that may change any of the above.

Definition of Healthiness or Competitiveness

As pointed out by many of the chapters, this is a principal source of confusion with the concept of competitiveness. The confusion seems to arise from two sources. First, a failure to specify the economic unit or level of aggregation of interest. This results in shifting across the nation, an industry, a firm, or a commodity, as the focus of analysis and an even more confusing specification of the definition (objective) for one of these, the statement of the problem for another and the proposal of a solution for still a third. Thus, whatever the definition of competitiveness, it should be specific to a level of aggregation or economic unit.

The second problem has been a failure to define competitiveness in terms of accomplishment of an agreed-upon objective or purpose for the specified economic unit. How can we hope to measure the degree of success (state of health) if we can't agree on the objective? An associated problem has been confusion of indicators or means with the objective.

For example, there seems to be reasonable agreement (in broad terms at least) that the objective of economic activity (competitiveness) at the national level is to maximize welfare (utility) of the population with all the appropriate considerations for equity among people and over time. If this is accepted as the goal, the problem is one of allocation of resources among uses and over time to provide the optimal mix (present and future) of goods and services. Trade is a means of maximizing the utility derivable from available resources. Yet, many discussions of competitiveness move directly from the above objective to trade performance as a measure of meeting the goal. This converts a means (and possible indicator) into an end. A large, small, increasing, or decreasing share of world trade is, per se, neither good nor bad. If maximization of national utility is accepted as the objective at the national level, what is the role of the industry or the firm? And, should its success (competitiveness) be judged on the contribution of the subunit to the national objective (e.g. value added) or some lower level objective—profit, return to capital, market share, etc.?

Identify Indicators of Healthiness (Competitiveness)

Most useful indicators would need to the observable, to be predictors of future change in healthiness and to indicate the illness, disease, injury, policy, economic behavior or change in environment that is the source of the problem. Few indicators are likely to be found that process all these characteristics. Identification depends on detailed understanding of the economy (body), its subsystems, environment, diseases and interactions. The key is to identify indicators that point to the source of the problem (the disease) and not just tell us we have a competitiveness problem.

Abbott and Bredahl, and Van Duren, Martin and Westgren, both propose frameworks for assessment and identification of determinants. Our received theories and models (both economic and managerial) provide the bases for indicators in both but are found incomplete and inadequate. Comparison of the two frameworks show the influence of the accepted definition on the kinds of indicators used.

At the national level, and given the earlier objective of economic activity, these might include keeping factor and product markets competitive, adopting policies that encourage private saving and investment, government investing in infrastructure and basic research, keeping trade barriers low and transparent, etc.

Identify Diseases and Treatments

Identify diseases and other conditions that negatively affect health. Here the analogy becomes more difficult but a few examples such as the Dutch Disease, short sightedness and protectionism come to mind. The difficulty is to distinguish between diseases and damage due to bad life style (violation of the kinds of rules discussed in the previous section). Confusion between the two may not be important so long as the nature of the illness is identified to the extent that both cause and effect are known and not just symptoms. Identification of the cause is necessary for development of a treatment.

Identify treatments specific to diseases. Treatments for economic diseases are most likely policies and other government action at the more aggregate levels and move toward strategic and tactical management action at industry and firm levels. Although some national level treatments such as "Buy America Campaign", consumers boycott and labor action are non-governmental. Treatments for economic diseases, at best, are likely to have side effects, some of which can be painful, and may not be predictable. Treatments should be targeted at the root cause of the illness, avoiding the indiscriminate prescription of broad spectrum antibiotics. For some economic illnesses the best treatment may be a change in lifestyle and/or environment. Also, as with humans, some patients die and one of the most difficult decisions is how long to keep the terminal patient on life support. Also aches and pains resulting from exercising and training should not be confused with arthritis in prescribing treatment.

Identify Environmental Changes

Be on the look out for changes in the environment that may change any of the above. Of course what is environment and what is internal to the economic unit depends on the level of concern—nation, sector, industry of firm. At lower levels of aggregation changes in environment may constitute treatments. However, the kind of environmental changes referred to here are those external to the economy. An example is the recent increase in concern about quality of the environment and sustainability of economic activity. Such change can result in modification of the societal conception of the objective of the economy (of what is competitiveness). Such changes may result in emergence of new economic diseases, require new indicatiors and new treatments.

Closing Observation

This has been a long way of suggesting that the relatively recent fixation on competitiveness by the public, policy makers and economists is proving counter productive. Even though Abbott and Bredahl tell us we are asking the wrong questions and suggests the right questions they return at the end to the objective of determining the patterns of trade. This is too narrow of a perspective and encourages the transformation of trades from a means of achieving national economic success into the objective of economic activity. Even worse, in the minds of many laymen, policy makers and more than a few economists the objective is not trade but exports.

If we could refocus (or rename) competitiveness on maximization of national product and away from trade competitiveness, I believe we as economists and policy makers would do more to improve the competitiveness of the U.S. economy (and the contribution of the food industry there to). We might even see the day when the following would occur:

> Industry X goes to Doctor Congress describing a competitiveness problem because he can't compete with low wage labor overseas, the industry asks whether Doctor Congress would again give him another prescription of quota protection and another bottle of subsidized credit, Doctor Congress will reply that research by the IATRC shows that such treatment would have a negative impact on national competitiveness, that his industry needs a change of environment, and that Congress will prescribe mobility capsules so that workers in industry X can more easily move to their better alternatives.

Research Implications of the Concept of Competitiveness

J. Bruce Bullock

I appreciate the opportunity to participate in this interesting conference. My assignment was to discuss the research implications of the papers and discussion during the conference from the perspective of a state agricultural experiment station administrator.

My first reaction to the set of papers I was sent was the same as Alex McCalla's. I did not want to see another definition of competitiveness. I was reminded of the story of the judge who was asked to define pornography. He said: "I cannot define pornography, but I know it when I see it." Being personally rather comfortable with Porter's definition of competitiveness, I thought we could surely know competitiveness when we saw it. However, after yesterday's discussion, I began to have doubts about our ability to even see competitiveness. Kelley White pointed out that the concept of competitiveness may be so nebulous as to be meaningless. I shutter to think that we may have created another term that generates as much confusion and unproductive discussion as the word "sustainability."

But, as Phil Abbott noted yesterday the administration at Purdue wants him to be working on this critical issue of "competitiveness." Why do college of agriculture administrators want research done on competitiveness? Simply because deans and directors base their justification for funding of their organizations on the need for U.S. agricultural producers to be competitive in increasingly global markets. The contributions of agricultural research to increased productivity are cited as evidence of the contributors of our college to the competitiveness of our state's agricultural producers.

As researchers interested in international trade, it should be nice to know that administrators are demanding the research you are doing. However, as an economist wearing an experiment station administrator's hat, I suggest that you have an educational challenge in front of you as well as some exciting research opportunities. Your educational challenge is to help research administrators and the production agriculture establishment understand the implications of what has been discussed here the past two days.

Table 17.1 summarizes the relative importance—great, some or little—of the determinants of competitiveness of each of the four economies of agriculture. For example, I judge natural resource endowments to be of great importance to competitiveness in primary commodities and of only some importance in differentiated primary products. But, I judge that it is of little importance to competitiveness in semi-processed and consumption-ready products. By contrast, product characteristics and non-price factors are judged to be of little importance to the first two economies, but of great importance to the last two.

TABLE 17.1 Importance of Selected Factors Determining for the Four Economies of Agriculture and Current Research Attention to Those Factors

Factors Determining Competitiveness	Production, Assembly, Transformation (Processing) and Final Distribution of:							
	Undifferentiated Primary Commodities		Differentiated Primary Products		Semi-processed Products		Consumption-Ready Products	
	Importance	Research Attention	Importance	Research Attention	Importance	Research Attention	Importance	Research Attention
Natural Resource Advantage, Factor Endowments	Great	High	Some	High	Little	Moderate	Little	Low
Cost Reducing Technology	Great	High	Some	High	Some	High	Some	Some
Human Capital and Managerial Expertise	Some	Moderate	Some	Moderate	Great	Low	Great	Low
Quality Enhancing Technology	Some	Moderate	Some	Moderate	Great	Moderate	Great	Low
Product Characteristics and Non-price Factors	Little	Low	Some	Low	Great	Low	Great	Low
Firm Strategy	Little	Low	Great	Low	Great	Low	Great	Some
Industry Structure Input Supply, Marketing and Distribution Channels	Some	High	Great	Low	Great	Low	Great	Low
Infrastructure	Great	High	Some	Moderate	Great	Low	Great	Low
Regulatory Environment and Trade Policies	Great, but declining	High	Varies greatly	High	Varies greatly	Low	Varies greatly	Low
Value Added	Little		Some		High		Highest	

The table also summarizes my judgement of the attention paid to, and the resources expended on, each determinant for each of the four economies at U.S. agricultural experiment stations - high, moderate and low. For example, cost-reducing (yield-increasing) technology receives a high level of research attention for primary commodities and products, a moderate level for semi-processed products, and little attention for consumption-ready products.

The contrast of the allocation of research efforts with the importance of each competitiveness factor is made by comparing the first and second column for each of the four economies. For example, factor endowments are of great importance for competitiveness in undifferentiated primary products and it receives a high level of research attention. That's good if you want to be competitive in undifferentiated primary commodity markets. Non-price factors are judged to be of great importance for competitiveness in semi-processed and consumption-ready products. But, the research attention is judged too low. That's bad if you want to be competitive in high-value added markets. A misallocation of resources is found where the importance is great but the research effort is low, or vice versa. The number of great importance / low research effort pairs in the last two columns is startling. And those are the economic activities purported to add-value to agricultural production.

The Old Research Paradigm

You need to recognize that most deans, directors and commodity group leaders are basing their views of competitiveness on a different paradigm than that discussed here. Their paradigm is that: "U.S. agriculture is the most efficient production machine in the world. We can compete with anyone as long as we have a level playing field." They will then top this statement off with a reference to how rapidly the world's population is growing relative to growth in production of agricultural products and, hence, our need to expand research in improving the efficiency of producing basic agricultural commodities. These deans and commodity group leaders then expect you to do research that searches for and exposes the bumps in the "non-level" playing field of international trade.

There is also an important second dimension to the agricultural establishment's paradigm. They view the world from a commodity orientation. They think in terms of #2 yellow corn, soybeans, #1 230 pound slaughter hogs, etc. They have not made the transition from commodities to food products. In terms of Table 17.1, they are aware of only the first two columns referring to undifferentiated primary commodities and differentiated primary products. There is a lot of rhetoric about the need to expand value-added exports that is perhaps reflected in column 3 of the table. But, column 4 is not included in the existing agricultural trade paradigm of traditional agricultural groups.

If you, as trade researchers, respond to this traditional paradigm in setting the research agenda, then you will continue to do research related to measuring and comparing labor productivity, capital productivity, etc. You will also continue to direct your trade research towards questions related to the world wheat market, the world soybean market, the world beef market, the world pork market, etc. On the other hand, if your research agenda is to reflect the new paradigm it will have a substantially different focus.

The New Research Paradigm

The new paradigm I heard the past two days revolves around the following observations. Countries do not compete at the macro level — in fact countries do not compete at all. Rather, the competition for trade occurs at the product and/or service level. Trade increasingly occurs in food products not commodities. World trade is demand driven not supply driven. There are large numbers of unique niche markets in the world.

Perhaps the most significant words in the title of this conference are food markets rather than the word competitiveness, which has been the focus of our discussion. When we apply this new (and still evolving) paradigm and focus on food trade and niche markets, new research questions and opportunities begin to emerge. We replace terms like world wheat market, world pork market and world beef market with terms like the U.S. pasta market, the Japanese pork loin market, and the German beef jerky market.

The significance of this new paradigm is being intensified by the technology products being developed by biotechnology research. In some cases, biotechnology will significantly overcome what have historically been viewed as "unfavorable" factor endowments of a country. In other cases, biotechnology may intensify the positive aspects of the natural endowments of another country. The results of biotechnology research will decompose commodity markets into product markets reflecting products with a different variability in product quality and generate the opportunity to produce products like pork loins that have size variations between portions that are only a small fraction of what we observe today. All these developments open opportunities for specified niche markets for products that replace commodity markets. The products of biotechnology research will further move us away from the Malthusian paradigm to a recognition that Engels Law is becoming the dominant factor affecting world trade in food products.

Thus, the research agenda for trade economists will be driven, to a large extent, by the paradigm directing the researcher's view of the world. Under the old paradigm of world markets in commodities, the research agenda does not change much. We will continue to search for and suggest remedies for bumps in the playing field such as trade barriers, etc. We will continue to focus on estimating commodity elasticities of demand and supply at rather aggregate levels. We will continue to do more of the same type of research we have done in the past.

However, the new paradigm suggests a substantially different research agenda. One might consider two types of research focus in the future. The first would be to describe and explain what happened to trade of a particular product over the past four years. One might describe this research as attempting to recognize "competitiveness" after the fact.

Another set of research will be an integral part of the process of facilitating the creation and use of competitive position. Can we identify and predict the necessary conditions for competitiveness in a particular food product market? Can we help U.S. business translate this information into trade?

This new research agenda will cause us to participate in (perhaps lead) efforts to improve the competitiveness of a particular product in specified markets. We will be involved in identifying market opportunities. Other research will involve determining why a particular product is not competitive in particular markets and evolving solution to the non-competitiveness problem. This research agenda will require expanded interaction between researchers and producers/exporters of food products. Research would focus on those areas identified as of great importance to competitiveness in the last two columns of Table 17.1.

As we focus on product markets in specific locations, I predict we, as economists, will increasingly become aware that institutions matter. The lack of market institutions in Eastern Europe and the former Soviet Union will provide some fascinating opportunities for research that can be useful in establishing the institutional network required to support well functioning markets for food products

Summary

In summary, it seems to me that much of the discussion the past couple of days could make a traditional trade researcher that has focused on commodity analysis at the rather macro level a bit uncomfortable. Much of the discussion suggests that more of the same research agenda is not longer appropriate. On the other hand, the research agenda suggested by the new (evolving) paradigm suggests some interesting and challenging research opportunities. I hasten to add that you might need to help educate your research administrators and traditional commodity group leaders as to why this new view of agricultural trade (i.e. food products rather than commodities) is appropriate. The new research agenda may well, at sometime, put you in the uncomfortable position pointed out by Kelley White's interesting analogy. You may be put in a position of diagnosing a particular patient as being terminally ill.

This has been a very interesting conference. You have been discussing some extremely important issues. I hope you will expand the discussion of this topic beyond the agricultural trade research community. Help the rest of us understand the tremendously significant implications of the new realities of world trade in food products.

What Did We Learn
from This Conference?

Alex F. McCalla

It has been an intense two days with 14 papers, plenty of discussion, and the preceding summary comments of my two distinguished colleagues. Ideally there should be little left to say. Further, your preferences are for home or libation, not another speaker. Yet I do have some things I want to share with you. The conference leaves me with an uneasy feeling that, while we have talked and listened a lot, we may not have communicated.

For starters a basic component of any discourse is a common language, which is based on universally agreed upon definitions. Over the past two days we have had almost as many definitions of competitiveness as we have had papers. There is clearly no agreed upon definition of what it is. One is tempted to ask—does it matter? It reminds me of many previous discussions of sustainability and farming systems research. Is competitiveness related to comparative advantage? Some say yes, while others argued they are separate and unrelated issues. Some say Heckscher-Ohlin neoclassical trade theory is no longer relevant and so on. No common definition emerged. The most we can conclude is that any definition must specify the *level* of economic activity under consideration, the *target* of the policy, and the *purpose* of the analysis.

If agreed upon definitions must precede constructive discourse, I am not sure how far we moved here. One got the impression that there are at least two camps at the meeting. I characterize them in their extremes to make my point about how wide a gap appeared to exist. One group I label neoclassical trade economists. The other is populated by people focused on markets and marketing in an industrial organization (I-O) framework.

Let me list some contrasts in their approaches to the topics at hand:
If one were to continue one could conclude that never the twain shall meet.

	Trade Economists	Marketers/I-O
1.	Theory based; abstract; deductive	Limited use of theory, sometimes *ad hoc*; inductive; talk of complex conceptual frameworks
2.	Puts premium on simplicity; the fewer variables the better	The more variables the better

3. Macro; general equilibrium;   Micro; firm, commodity or
    international, open economy;   industry based; nationalistic
    focus on sector, national,
    international markets

4. Price absolutely central in   Non-price elements are much
    terms of efficient resource   more important than product
    allocation—also macro prices—   price
    exchange rates, interest rates

5. Seeks generalization   Focus on differentiation

6. Quantitative models based   Quantitative *ad hocery*; focus on
    on theory with testable   niches; strong role for
    econometric properties   qualitative analysis

7. Policy variables central to   Policy at best is part of a vague
    the analysis   institutional environment

8. Strongly supply driven—   Demand driven, with a strong
    commodity oriented—focused   marketing orientation
    on inputs, products and
    relative costs

Were the divisions bridged here or was the fissure widened? At times it seemed like the latter.

Yet, clearly in my mind, both approaches have a role to play in addressing the central issue of what determines how well countries, industries, and products do in international markets. The intellectual challenge is to build critical linkages between the two approaches. Will research help? I think the answer is probably yes. For example, research should help us lay to rest the silly proposition that high value (value-added) exports are always to be preferred by all countries.

In the remainder of my time, I suggest a random set of critical research issues that seemed to emerge here where both camps have something to contribute.

Understanding the role of multi-national firms. Why do food processing firms seem to prefer direct foreign investment to trade? What is their role in marketing? Does the nationality of headquarters really matter? Do they exercise market power? Is there evidence of interdependence with governments?

Learning how to include "non-economic (read non-price) variables" in demand analysis. Some papers were rich in detail about product characteristics, but gave us little insight into how these would be factored into more aggregate forecasting and analysis models.

The role of case studies vis-a-vis trade models. Case studies frequently are addressed to the firm, product, sector, or country. They are rich in institutional, product, and technical detail. But alone they have limited global value unless their results can be aggregated and generalized. Working together could help to make the best joint use of global models and detailed case studies.

Most models of trade liberalization are focused on bulk products. *What do we know about the impact of liberalization of agriculture and other areas on trade in manufactured and processed food product?.* There are two related questions. Have previous reductions in trade barriers in manufactured products spilled over into processed food markets? Second, as Maury Bredahl forcefully pointed out, understanding marketing channels and non-price characteristics are critical to the analysis of trade liberalization. Traditional trade models are notoriously unreliable in understanding and predicting bilateral trade flows.

Understanding the similarities and differences between firm welfare and national welfare. In whose interest is the trade occurring? As Li'l Abner used to say, "What's good for General Bullmoose (Motors) is not necessarily good for the nation." Clearly research requires a clear delineation of whose welfare function we are trying to optimize.

This is only a partial list, but I believe it demonstrates that the needs are large. What then is the bottom line? Clearly there are great intellectual challenges as we try to confront traditional trade models with reality. Recent developments in so called "new trade theory" have attempted to modify the theory to more closely fit reality. But the relevance of this approach to agricultural and food trade has yet to be demonstrated. But perhaps we should be careful about tinkering with long received theory. Theory by definition is focused on developing the simplest model to explain broad tendencies as for example is the Theory of Comparative Advantage. Therefore don't ask theory to explain what it was not intended to explain, e.g. changes in U.S. pork exports to Japan. Further in criticizing theory we must be careful not to throw the baby out with the bath water. Talk about repealing the Theory of Comparative Advantage is silly. Without an organizing framework, micro-investigation becomes time-dated *ad hocery.*

What we need to do is find ways to allow theory-based trade economists to pool their strengths with commodity and firm specialists. This latter groups' knowledge of products and markets is absolutely critical to understanding linkages between firms, products, and markets, both national and international. However, his challenge will not be easily met. Maybe this meeting was a first step. Lets hope it wasn't backward.

About the Contributors

Philip C. Abbott, Professor, Department of Agricultural Economics, Purdue University

John W. Allen, Professor, Marketing and Logistics, Michigan State University

Ken Ash, Chief, Competitiveness Division, Agriculture Canada

Maury E. Bredahl, Professor, Department of Agricultural Economics, Missouri University

Lars Brink, Senior Economist, Competitiveness Division, Agriculture Canada

J. Bruce Bullock, Associate Dean, College of Agriculture, Missouri University

Wilfred J. Ethier, Professor, Department of Economics, University of Pennsylvania

Brad Gehrke, Graduate Research Associate, Department of Agricultural Economics, Missouri University

Charles R. Handy, Economics Research Service, U.S.D.A.

Tim Hazledine, Professor, Department of Economics, University of Auckland

Dennis R. Henderson, Professor Emeritus, Department of Agricultural Economics and Rural Sociology, The Ohio State University

Nicholas G. Kalaitzandonakes, Assistant Professor, Department of Agricultural Economics, Missouri University

Ralph G. Lattimore, Professor, Department of Economics and Marketing, Lincoln College

Lawrence J. Lau, Prefessor, Department of Economics, Stanford University

John E. Lee, Jr., Administrator, Economics Research Service, U.S.D.A.

Larry Martin, Director, University of Guelph, G. Morris Policy Center

Alex F. McCalla, Professor, Department of Agricultural Economics, University of California, Davis

Steve McCorriston, Professor, Department of Economics, University of Exeter

Stephen MacDonald, Agricultural Economist, Economics Research Service, U.S.D.A.

William M. Miner, Senior Trade Researcher, Center for Trade Policy and Law

Thomas Pierson, Professor, Department of Marketing, Michigan State University

Michael R. Reed, Professor, Department of Agricultural Economics, University of Kentucky

Ian Sheldon, Associate Professor, Agricultural Economics, and Rural Sociology, The Ohio State University

Erna van Duren, Associate Professor, Department of Agricultural Economics, Guelph University

Aage Walter-Jørgensen, Professor, Institute of Agricultural Economics, Copenhagen, Denmark

Randall Westgren, Chairman, Department of Agricultural Economics, McGill University

T. Kelley White, Director, Policy Analysis Division, UNFAO

About the Book and Editors

The successful completion of the GATT negotiations and the North American Free Trade Agreement and the completion of the EC Internal Market mean that food and agricultural sectors must become internationally competitive. Firms, farm organizations, and governments are seeking to identify strategies and public policies that will increase their competitiveness. This book draws together papers that define and evaluate the concepts that underlie firm, sector, and national competitiveness. The book is divided into four parts:

1. Conceptual Foundations and Assessments from Business Strategies
2. Conceptual Foundations and Assessments from Trade and Macroeconomic Theory
3. Assessments of the Competitiveness of National Food Sectors
4. Commentary and Implications

In the first part, the business school approach to competitiveness is used to develop a conceptual framework and then key elements are used in empirical assessments of competitiveness. The second part draws a conceptual framework from international trade theory which serves as the basis for empirical applications to competitiveness assessment. The third part reports a number of assessments of the competitiveness of the U.S., Canadian, New Zealand, and Danish food and agricultural sectors. The role of the determinants of competitiveness, of foreign direct investment, and of trade and agricultural policies are empirically investigated. In the final part, a panel of eminent economists take very different views of the usefulness of competitiveness as a paradigm for guiding a research agenda and as basis for government policy formulation.

Maury E. Bredahl is professor of agricultural economics and director of the Center for International Trade Expansion at the University of Missouri. **Philip C. Abbott** is professor of agricultural economics at Purdue University. **Michael R. Reed** is professor of agricultural economics and director of the Center for Agricultural Export Development at the University of Kentucky.

Index

Printed and bound by CPI Group (UK) Ltd, Croydon, CR0 4YY

28/10/2024

01780298-0006